The Minerals, Metals & Materials Series

Bowen Li · Shefford P. Baker · Huazhang Zhai ·
Sergio Neves Monteiro · Rajiv Soman ·
Faqin Dong · Jinhong Li · Ruigang Wang
Editors

Advances in Powder and Ceramic Materials Science

TMS

Springer

Editors
Bowen Li
Michigan Technological University
Houghton, MI, USA

Huazhang Zhai
Beijing Institute of Technology
Beijing, China

Rajiv Soman
Eurofins EAG Materials Science
Liverpool, NY, USA

Jinhong Li
China University of Geosciences
Beijing, China

Shefford P. Baker
Cornell University
Ithaca, NY, USA

Sergio Neves Monteiro
Military Institute of Engineering
Rio de Janeiro, Brazil

Faqin Dong
Southwest University of Science
and Technology
Mianyang, China

Ruigang Wang
The University of Alabama
Tuscaloosa, AL, USA

ISSN 2367-1181 ISSN 2367-1696 (electronic)
The Minerals, Metals & Materials Series
ISBN 978-3-030-36551-6 ISBN 978-3-030-36552-3 (eBook)
https://doi.org/10.1007/978-3-030-36552-3

Cover illustration:
From Chapter "Toughening Mechanism of ZTA–TiC–Fe Ceramic Materials Produced by High-Gravity Combustion Synthesis", Hongwei Zhao et al, Fig. 4 Toughening mechanisms in ZTA–TiC–Fe ceramic material: a ZrO_2 fibers toughening; b ZrO2 transformation toughening; c toughening by ductile metal phase. https://doi.org/10.1007/978-3-030-36552-3_4

This Springer imprint is published by the registered company Springer Nature Switzerland AG
The registered company address is: Gewerbestrasse 11, 6330 Cham, Switzerland

Preface

Ceramic materials science covers the science and technology of creating objects from inorganic, nonmetallic materials, and includes design, synthesis, and fabrication of ceramics, glasses, advanced concretes, and ceramic–metal composites. Metallurgy and ceramics always share similar technical principals and methodologies in studying processes and structures of diverse materials from powder preparation, forming, and sintering, until product finishing. In recent years, the hybrids of ceramic materials and metallic materials have received plenty of interdisciplinary inspirations and achievements in material processes and functional applications including ionic batteries, catalysis, energy storage, superconductors, semiconductor, filtrations, and additive manufacturing.

To promote interdisciplinary studies, The Minerals, Metals & Materials Society (TMS) Materials Characterization Committee and Powder Materials Committee co-sponsored the symposium "Advances in Powder and Ceramic Material Science" at the 2020 TMS Annual Meeting. This symposium focused on the advances of powder and ceramic materials in the fundamental research, technology development, and industrial applications. The subjects of the symposium include synthesis, characterization, modeling, and simulation of powder and ceramic materials; design and control of ceramic microstructure and properties; ceramic powders and processing; surface treatment and thin films, membranes, and coatings of ceramics; hybrid systems of ceramic, metal, and/or polymer composites; and metallurgical byproducts for ceramic manufacturing.

This proceedings volume includes 16 reviewed manuscripts of original research. The manuscripts were invited or contributed by the researchers from the fields of powder processing, materials science, engineering, metallurgy, physics, manufacturing, and applications. This book provides the authors' up-to-date achievements in powder and ceramic materials. It would be a useful reference for academic and industry readers.

The editors would like to acknowledge the authors for their contributions to the manuscripts, the TMS Materials Characterization Committee and Powder Materials Committee for their support, and Springer for publishing this volume.

Bowen Li
Lead Organizer

Contents

Part IV Ceramic-Based Composite Materials

Part V Poster Session

About the Editors

Bowen Li is a Research Professor in the Department of Materials Science and Engineering and Institute of Materials Processing at Michigan Technological University. His research interests include materials characterization and analysis, metals extraction, ceramic process, antimicrobial additives and surface treatment, porous materials, applied mineralogy, and solid waste reuse. He has published more than 120 technical papers in peer-reviewed journals and conference proceedings, authored/co-authored 3 books, and edited/co-edited 8 books. He also holds 15 patents and has delivered more than 30 invited technical talks.

Dr. Li received a Ph.D. degree in Mineralogy and Petrology from China University of Geosciences Beijing in 1998, and a Ph.D. degree in Materials Science and Engineering from Michigan Technological University in 2008. He has been an active member in The Minerals, Metals & Materials Society (TMS), Society for Mining, Metallurgy & Exploration (SME), and China Ceramic Society. At TMS, he currently serves as the Past Chair of the Materials Characterization Committee and as a member of Powder Materials Committee and Biomaterials Committee. He also has served as an EPD Award Committee member, a *JOM* Subject Advisor, and a Key Reader for *Metallurgical and Materials Transactions A*. He has been Organizer/Co-organizer of a number of international symposia and sessions. He also served as an editorial board member of the *Journal of Minerals and Materials Characterization and Engineering, Reviews on Advanced Materials Science,* and *FUTO Journal Series.*

Shefford P. Baker received his Ph.D. in Materials Science and Engineering at Stanford University and was a staff scientist at the Max-Planck-Institut für Metallforschung in Stuttgart before joining the faculty in Materials Science and Engineering at Cornell University. He has been Visiting Professor at the Faculté des Sciences et Techniques de Saint Jérome, Université Paul Cézanne, Marseille, was President of the Materials Research Society, and has received several teaching awards. He is a member of TMS, MRS, and ACerS. His research focuses on the mechanical properties of materials, particularly those having critical length scales in the nanometer regime, with a focus on plastic deformation mechanisms and the interactions among stresses, structure, and mechanical performance. His interests include relating processing to structure, defects, and stresses in thin films including texture and phase transformations; structure, composition, and plastic deformation in silicate glasses; defects and nanomechanical behavior in biogenic minerals and their synthetic analogs; nanocontact mechanics; and structure and deformation in advanced metals.

Huazhang Zhai is an Associate Professor in the School of Materials Science & Engineering and the Director of Materials Chemistry & Physics Laboratory at Beijing Institute of Technology, China. He earned his B.S. degree in Chemistry from Anhui Normal University in 1995 and Ph.D. degree in Materials Science and Engineering from Tsinghua University in 2003. He was a Research Associate Professor in the Department of Applied Physics at Hong Kong Polytechnic University in 2009 and a Visiting Scholar in the Department of Aerospace Mechanics at Princeton University in 2014.

Dr. Zhai's primary research interests include advanced ceramics, environmental nanomaterials, and novel energy materials. He has led two research projects funded by the National Natural Science

Foundation of China and a number of projects from other sources. He is the author of more than 50 peer-reviewed journal articles. He is the member of China Materials Research Society Young Scientist Council.

Sergio Neves Monteiro graduated as a metallurgical engineer (1966) at the Federal University of Rio de Janeiro (UFRJ). He received his M.Sc. (1967) and Ph. D. (1972) from the University of Florida, followed by a 1975 course in energy at the Brazilian War College, and a postdoctorate (1976) at the University of Stuttgart. In 1968, he joined the Metallurgy Department of UFRJ as Full Professor of the postgraduation program in engineering (COPPE). He was elected as Head of Department (1978), Coordinator of COPPE (1982), and UnderRector for Research (1983), and was invited as UnderSecretary of Science for the State of Rio de Janeiro (1985) and Under-Secretary of College Education for the Federal Government (1989). He retired in 1993 from the UFRJ and joined the State University of North Rio de Janeiro (UENF), from where he retired in 2012. He is now Professor at the Military Institute of Engineering (IME), Rio de Janeiro, and has published more than 1,200 articles in journals and conference proceedings and has been honored with several awards including the ASM Fellowship. He is top researcher (1A) of the Brazilian Council for Scientific and Technological Development (CNPq) and Top Scientist of State of Rio de Janeiro (FAPERJ). He was President of the Superior Council of the State of Rio de Janeiro Research Foundation, FAPERJ (2012), and currently is coordinator of the Engineering Area of this foundation. He also has served as president of the Brazilian Association for Metallurgy, Materials and Mining (ABM, 2017–2019), a consultant for the main Brazilian R&D agencies, and a member of the editorial board of five international journals as well as associate editor of the *Journal of Materials Research and Technology*.

Rajiv Soman currently serves as Director, Purity Survey Analysis, Materials Sciences Division at Eurofins EAG Laboratories, USA. He has over 30 years of professional experience in analytical chemistry and materials sciences. He earned a doctorate in Analytical Chemistry from Northeastern University, Boston. He received his B.Sc. (Chemistry—*Principal*; Physics—*Subsidiary*) with Honours, from Bombay University, India, and M.Sc. in Applied Chemistry from the Faculty of Technology & Engineering, Maharaja Sayajirao University of Baroda, India. He commenced his professional career as an Advanced Analytical Chemist in the Engineering Materials Technology Laboratories of General Electric Aircraft Engines. Prior to joining EAG Laboratories, he served as Professor (Full) of Chemical Engineering, Chemistry, and Chemical Technology, and served as a faculty member for 20 years. He has received numerous awards for excellence in teaching and twice has been listed in *Who's Who Among America's Teachers*.

Dr. Soman's research interests are in the areas of atomic and mass spectrometry, with emphasis on trace element determination and chemical speciation in a wide range of sample matrices. He was an invited guest scientist at the prestigious research institute, Forschungszentrum Jülich, Germany, where he conducted research in elemental mass spectrometry. He has co-authored several publications in international journals, and has made numerous presentations at national and international conferences. He holds two U.S. patents.

Dr. Soman has been a member of Society for Applied Spectroscopy (SAS) and American Chemical Society (ACS) since 1986 and has served in numerous leadership positions in the societies. He was an invited panel member for the American Chemical Society's National Committee on *Preparing for the Workforce 2020*. He is also a member of ASM International, ASTM, and TMS, where he serves as a member of the Materials Characterization Committee.

Faqin Dong is a Professor at Southwest University of Science and Technology (SWUST), Mianyang, China, and Director of the Key Laboratory of Solid Waste Treatment and Resource Recycle of the Ministry of Education, China. His research interest mainly focuses on mineral materials, environmental materials, particuology, microbe–material interaction, and environmental engineering. Dr. Dong has led more than 20 research projects sponsored by the National Natural Science Fund of China, the Ministry of Education, National Bureau of Building Materials Industry, and the National 863 and 973 Plans. He has published more than 160 research papers in peer-reviewed journals and proceedings, and he holds 13 patents. He has authored four books including *Applied Mineralogy* and *Applied Mineralogy of Fibrous Brucite*, and edited four books including *Eco-functional Elementary Materials and Integration Technology of Composite Building Materials* and *Research on Environmental Mineralogy and Environmental Medicine*. He is the organizer of a number of national/international conferences. He is the recipient of several national/provincial science and technology awards and educational awards.

Dr. Dong earned his B.S. degree in Exploration and Geology of Nonmetallic Minerals from Sichuan Industrial Institute of Building Materials, China in 1985 and Ph.D. degree in Mineralogy from China University of Geosciences (Wuhan), China in 1992. He was a Visiting Scholar at Laurentian University, Canada, in 1992–1993.

Jinhong Li is a Professor and the Associate Dean of the School of Materials Science and Engineering at China University of Geosciences, Beijing. He received his Ph.D. degree from China University of Geosciences, Beijing in 2007. His research interests include mineral energy storage materials, porous ceramics, and utilization of mineral and solid waste resources. He has undertaken more than 20 research projects; published more than 70 papers in internationally renowned journals such as *Journal of Materials Chemistry A, Energy, Applied Energy, Journal of the European Ceramic Society, Journal of Power Sources, Solar Energy Materials and Solar Cells, and Materials*

& Design; co-authored one book; and served as a reviewer for a number of international journals. He also holds four patents. He was the winner of the New Century Talent Program from the Ministry of Education of China in 2008, the National Excellent Doctoral Dissertation in 2010, and was awarded the 12th Youth Geological Science Silver Hammer Award in 2011. He was a Visiting Scholar in the Department of Materials Science and Engineering at Pennsylvania State University, USA, from September 2011 to September 2012.

Ruigang Wang received his Ph.D. in Materials Science and Engineering from Arizona State University in 2007. His Ph.D. dissertation work was on in situ environmental transmission electron microscopy study. He then did a postdoc at the Materials Sciences Division, Lawrence Berkeley National Lab, working on intermediate-temperature fuel cells and Li-ion battery materials. Currently, he is an associate professor in the Department of Metallurgical and Materials Engineering at The University of Alabama, Tuscaloosa. His research focuses on the synthesis and processing–structure–property relationship study of oxides, emission control catalysts, catalyst support materials, energy conversion/storage materials (fuel cell and battery, etc.), and high-temperature ceramic processing.

Part I
Structure Design and Processing

Sintering Process and Characteristics of Glass-Ceramics from Coal Fly Ash

Li Zeng, Hongjuan Sun, Tongjiang Peng and Wenmiao Zheng

Abstract In this study, glass-ceramics was prepared with coal fly ash by direct sintering to solve the problem of resource utilization of coal fly ash. The effects of the sintering temperature on volume density, water absorption, and open and closed porosity from 1000 to 1200 °C were investigated. The sintering process of glass-ceramics was also investigated by XRD and SEM. The data indicate that the sintering process was accomplished by dissolving the solid amorphous phases, hematite, and portion quartz of coal fly ash to form anorthite. During the process, the open porosity of the material gradually tended to closure. The glass-ceramics fired at 1180 °C for 30 min featured good strengths, with compressive strength of 122 MP and bending strength of 34 MP. The obtained relatively low density of 1.94 g cm^{-3} and moderate water absorption of about 1% would make these materials suitable for lightweight construction tiles. Besides, owing to the abundance of coal fly ash produced from thermal power plants, the present technology should be suitable for large-scale manufacturing of glass-ceramics, with economic benefits and possible solutions to environmental and wastes disposal concerns.

Keywords Coal fly ash · Glass-ceramics · Anorthite · Sinter process · Lightweight construction tiles

---

L. Zeng · H. Sun (✉) · T. Peng · W. Zheng
Key Laboratory of Solid Waste Treatment and Resource Recycle,
Southwest University of Science and Technology, Ministry of Education,
Mianyang 621010, Sichuan, People's Republic of China
e-mail: sunhongjuan@swust.edu.cn

L. Zeng
School of Architecture and Civil Engineering, Chengdu University, Chengdu 610106,
Sichuan, People's Republic of China

B. Li et al. (eds.), *Advances in Powder and Ceramic Materials Science*,
The Minerals, Metals & Materials Series,
https://doi.org/10.1007/978-3-030-36552-3_1

Introduction

China is one of the few countries that use coal as major energy source, accounting for about 63% of China's primary energy consumption. Coal would still be the primary energy source in the near future. Currently, large amounts of fly ash have been produced from thermal power plants, consuming about 45% of total coal outcome each year [1]. The emissions of coal fly ash exceed 580 million tons per year in China. On the other hand, only 70% of China's coal fly ash is taken into account while the remaining causes serious environmental pollution, including haze and smog issued from random accumulation [2, 3].

Oxides like silica, alumina, calcium oxide, and iron oxide in coal fly ash are considered as low-cost raw materials for preparing glass-ceramics. As a result, various types of glass-ceramics are prepared from fly ash by controlling nucleation and crystallization process through traditional melting and sintering methods with addition of natural wastes and raw materials, such as shell, tailings, waste glasses, red mud, and silica [4–7]. It has to be kept in mind that direct sintering is similar to traditional sintering but favored by scholars due to its low energy consumption.

Sintered samples made from direct sintering are referred as glass-ceramics owing to phase transformation during the sintering process [8–10]. In fact, coal fly ash generated at high temperature followed by sudden cooling contains more than 50% amorphous glassy phase [11]. During direct sintering, amorphous glassy phase will melt and react with each other to form aluminosilicate crystals. Therefore, coal fly ash can be used to prepare glass-ceramics by direct sintering process. For instance, Lu et al. prepared glass-ceramics by direct sintering from mixtures of magnesia, waste glass, and fly ash, and investigated the influence of particle size on sinterability, crystallization kinetics, and flexural strength of the resulting glass-ceramics [12, 13], while the amounts of coal fly ash did not exceed 50%. Due to the large amount of coal ash produced each year, producing glass-ceramics is a useful approach for the consumption of fly ash.

In this work, direct sintering was used to prepare sintered glass-ceramics from coal fly ash. The sintering process and characterization of glass-ceramics sintered at different temperatures were studied by XRD, SEM, and Archimedes' method. The sintering activation energies were calculated. These investigations are helpful to understand the application of coal fly ash in the preparation of glass-ceramics.

Experimental

Coal fly ash used in this study was taken from a power plant in Mianyang, Sichuan, China, and labeled as CAF. The chemical compositions of CFA were measured by X-ray fluorescence (XRF, PANalytical), and the results are listed in Table 1. The CFA contained large amounts of alumina and silica with small amounts of calcium and iron. The loss of ignition (LOI) of coal combustion residues was

Table 1 Chemical compositions of CFA (wt%)

Composition	SiO_2	Al_2O_3	CaO	Fe_2O_3	K_2O	SO_3	TiO_2	Na_2O	MgO	P_2O_5	SrO	BaO	LOI
Fly ash	53.31	26.55	5.88	4.41	2.42	1.42	1.07	0.60	0.52	0.47	0.12	0.10	3.02

estimated to 3.02%. The particle size of CFA was analyzed by a laser particle size analyzer (Zetasizer Nano Zs90, Fig. 1). The CFA powders showed broad particle size distribution ranging from 1 to 100 um, with a mean size of about 12.62 um.

The samples were formed into disks of 25 mm in diameter from a mixture of 3 g CFA with 5 wt% additional PVA (polyvinyl alcohol) at pressure of 10 MPa applied for 30 s. The PVA was used as a binder for the disks for increasing the green strength and decreasing the initial porosity. The samples were sintered in an electric box furnace at 1000, 1100, 1150, 1160, 1170, 1180, 1190, and 1200 °C, respectively, for 10, 20, 30, 40, 50, and 60 min with heating rate of 10 °C/min. After sintering, the samples were naturally cooled down to room temperature. The obtained sintered specimens were labeled as CFA-sinter temperature-holding time. The schematic representation of glass-ceramics is shown in Fig. 2. Water absorption (W) was measured using standard tests in boiling distilled water for 1 h. The bulk density (ρ_b) and open porosity (ρ_o) were measured by the Archimedes method. The true densities (ρ_t) were measured by gas (Ar) pycnometer. The results were used to estimate the total (ρ_a) and closed (ρ_c) porosities using Eqs. (1) and (2).

$$\rho_a = 1 - \frac{\rho_b}{\rho_t} \times 100\% \tag{1}$$

$$\rho_c = \rho_a - \rho_o \times 100\% \tag{2}$$

With respect to other properties, three-point bending strength was measured using rectangular bars of 5 × 10 × 60 mm^3 at 0.5 mm/min displacement. The

Fig. 1 Particle size of CFA

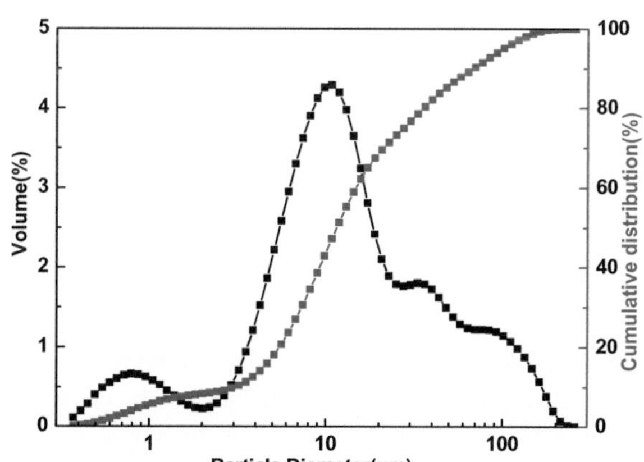

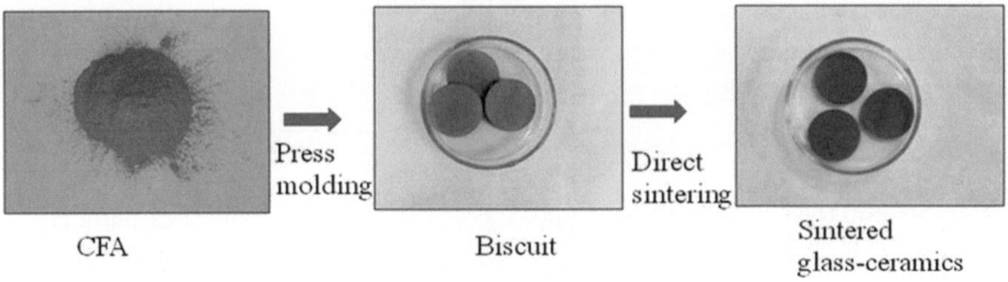

Fig. 2 Schematic representation for glass-ceramics

sintering kinetics of glass-ceramics was based on bulk density. The empirical Eqs. (3) and (4) were employed to calculate the sintering activation energies.

$$D = K\log t + C \tag{3}$$

$$K = A\exp\left(-\frac{Q}{RT}\right), \tag{4}$$

where D is the bulk density, C is a characteristic constant of the powders, K is the reaction rate constant, t is sintering time, Q is the activation energy, R is gas constant, T is the absolute temperature, and A is a constant [14, 15].

The crystalline structures of glass-ceramics were determined by X-ray diffraction (XRD, PANalytical, 2θ range 3°–80°, step 0.03°). The microstructures of glass-ceramics were observed using scanning electron microscopy (SEM, Ultra 55) operating at 15 kV at magnifying multiple ranges of 1000–500. The samples were crushed and gold sputtered, and both surface and fractured surface were then analyzed.

Results and Discussion

Figure 3 showed that the sintering process started at 1000 °C while overfiring was observed at 1190–1200 °C. The glass-ceramics CFA1180-60 showed the minimum water absorption of 4.4% (Fig. 3a), which rose to higher water absorption due to overfiring. On the other hand, bulk density revealed an opposite trend, with maximum bulk density of 1.86 g/m^3 at the same temperature. Also, bulk density declined due to overfiring. The same tendency can be seen in Fig. 3b: a regular decrease in open porosity was observed as a function of temperature, which still affected by about 8% of open porosity after sintering at 1180 °C. A regular increase in closed porosity was observed at temperatures below 1180 °C. The plot in Fig. 3b depicted a typical sintering behavior, with decreased open porosity and increased permanence of very few closed porosities [16]. The 4.4% water absorption estimated for CFA-1180-60 was caused by the 8% open porosity. After sintering at

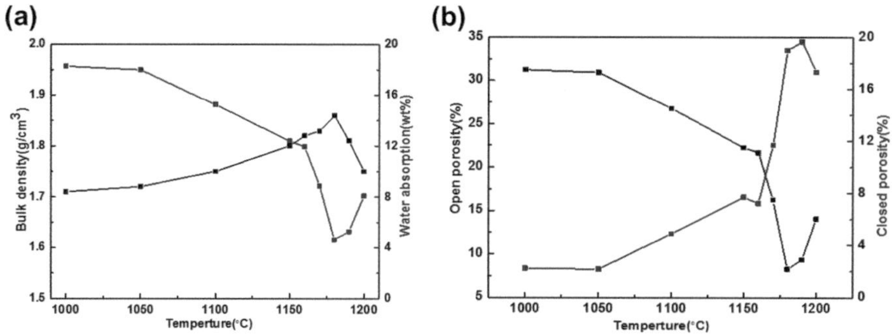

Fig. 3 Property of sintered samples at different temperatures **a** Bulk density and water absorption. **b** Open and closed porosities

1180 °C, the outer surface of each sample showed open porosity due to overfiring, which can cause deformation. The water absorption showed an increasing trend and bulk density declined. In general, the sintering process stared at 1000 °C and as sintering progressed, the open porosity gradually transformed into closed or vanished porosity. Also, water absorption of the samples gradually reduced and bulk density rose due to increased viscous flow caused by melting material in CFA.

To meet suitability for applications in lightweight construction tiles with low water absorption, the effects of holding time at temperatures of 1160, 1170, and 1180 °C were investigated and the results are shown in Fig. 4. The increase in holding time led to similar trend as the increase in sintering temperature (Fig. 4a) since water absorption first declined and then increased due to overfiring. It will be noted that the sintering process of CFA1180 reached completion after 30 min and sintered samples CFA-1180-30 showed lowest water absorption values (1.00%). Interestingly, water absorption reached about 7% after 50 min at 1160 °C, 4% after 40 min at 1170 °C, and 1% after 30 min at 1180 °C. The latter could be attributed to the similar trends observed for open porosity (Fig. 4b).

Sintering was driven by the increased viscous flow caused by melting of non-crystalline and crystalline material in CFA. Hence, raising the holding time and especially sintering temperature may reduce the viscosity of the glassy phase causing collapse of structure and densification of sintered samples [15]. Especially densification the surface of the samples, because the surface is generally hotter than the rest for the electric furnace heating [17].

To explain the sintering process of glass-ceramics, the samples were analyzed by XRD and SEM and their sintering activation energies were determined at different sintering temperatures. Main phase transformations observed at different temperatures are gathered in Fig. 5. CFA contained mainly quartz, mullite, and hematite with amorphous metastable glassy phase forms. These data were consistent with those published in other literature [3, 18, 19]. The sample treated at 1000 °C demonstrated slight decrease in quartz and amorphous metastable glassy phases. Also, anorthite began to appear and hematite phase totally vanished. At 1100 and

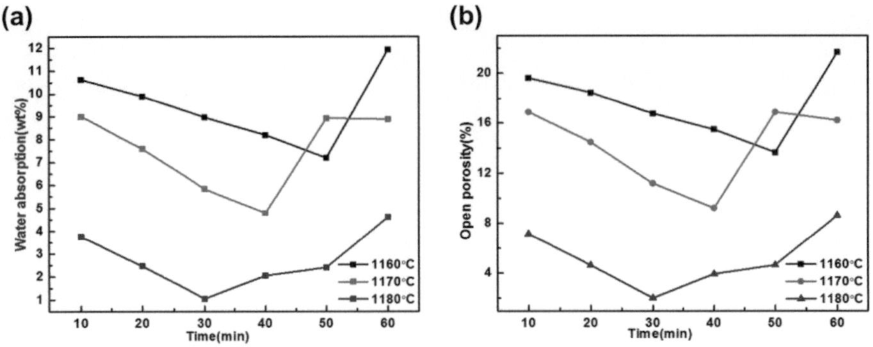

Fig. 4 Property of samples sintered at 1160, 1170, and 1180 °C with different times. **a** Water absorption. **b** Open porosity

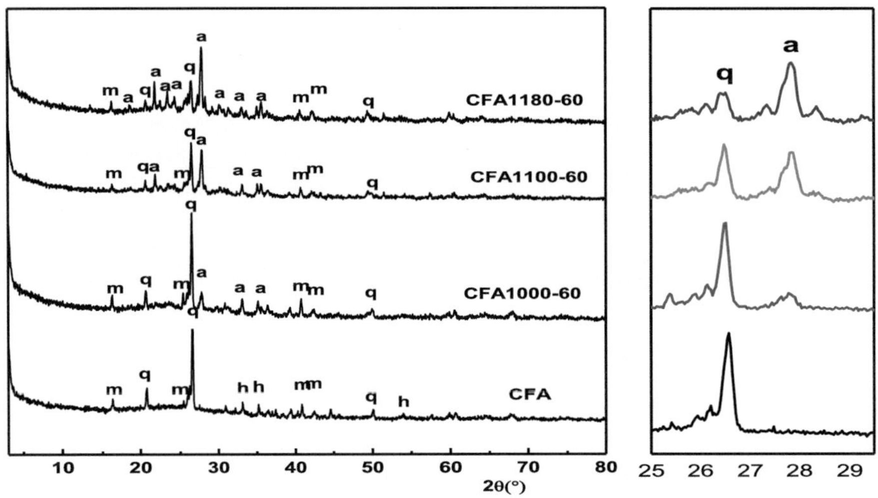

Fig. 5 XRD spectra of CFA and sintered samples after 60 min at different temperatures (q-quartz, m-mullite, a-anorthite, and h-hematite)

1180 °C, the decrease in quartz and amorphous metastable glassy phase became significant while anorthite amounts increased. In other words, quartz and mullite as residual minerals and anorthite were reconstructed from amorphous metastable glassy phase, hematite and quartz during the firing process from CFA. The diffuse diffraction background of CFA in XRD data was related to CaO content [20] and aluminum silicate glassy phase. The excess energy fixed in amorphous metastable glassy phase made them inclined to release excess energy by crystallization [21]. As sintering temperature rose, the XRD background levels and intensity of quartz decreased due to portion dissolution [22]. The intensities of hematite vanished and those of anorthite increased indicating less amorphous glassy phase and shift to anorthite. Qin et al. [23] reported that the synthesis of anorthite at low temperatures was challenging except for small particle size raw materials. The powder particle size analysis estimated d_{50} of CFA to 12.62 μm (Fig. 1). Hence, low-temperature fabrication of anorthite was successfully achieved by CFA at 1180 °C.

Figure 6 shows the SEM micrograph of both surface and fracture surface of sintered samples at different temperatures. The sintering densification process occurred as temperature increased. At 1000 °C, both the surface and fracture surface appeared uneven and granular (Fig. 6a, d). The increase in firing temperature to 1100 °C resulted in smoother surfaces, and fracture surface shown in Fig. 6b, e, indicated melting and sintering processes. As firing temperature further rose, porosity of fracture surfaces of sintered samples increased and showed changes from unconnected to interconnected shapes. Samples at low temperatures illustrated unconnected and irregular sized pores (5–10 um) caused by porosity and incomplete reaction of coal ash itself (Fig. 6d, e). By contrast, samples at relatively high temperatures depicted larger and interconnected sized pores of 5–30 um (Fig. 6f). Here, porosity was determined by sinterability [19], which should be ascribed to the lower viscosity of liquid phase caused by high temperatures of the densification process. By contrast, surface porosity of sintered samples greatly decreased as firing temperature increased consistent with water absorption data (Fig. 3a). Figure 6 showed typical sintering behavior, with open porosity of surface decreased. Meanwhile, the closed porosity of fracture surface increased.

Equation (3) revealed that the reaction rate constant (k) was determined by the slope of the plot: D (bulk density) versus log*t* (logarithmic scale of sintering time shown in Fig. 7a). Figure 4 suggested that the sintering process reached completion after 30 min for CFA1180, 40 min for CFA1170, and 50 min for CFA1160. The plot D versus log*t* was consistent with Fig. 4. Also, the sintering activation energy can be calculated from the slop: *ln* K versus 1/T graphs of glass-ceramics shown in Fig. 7b. The sintering activation energy of glass-ceramics was estimated to 127 kJ/mol, which was lower than that reported by Lu et al. [12] for glass-ceramics sintered from fine powders. This may be attributed to less time used during the sintering process. In other words, coal fly ash could easily be sintered not only for high special surface energy of fine powder but also for time during sintering process.

To acquire sintered samples suitable for lightweight construction tiles with limited water absorption, CFA1180-30 with least water absorption was produced. The physical and mechanical properties of the sample are listed in Table 2. The 1% water absorption obtained for CFA 1180-30 was caused by absorption from sample's side without any dense surface layer [17]. The density of CFA 1180-30 was estimated to only 1.94 g cm^{-3}, much was lower than those of other materials reported in the literature [6, 12]. This made it suitable for applications in lightweight construction tiles, especially for tiles that can be placed vertically for coverage of internal walls and manufacturing of ventilated facades. The compressive and bending strengths were issued from the compactness of the samples, while values of glass-ceramics were quite lower than those prepared by Yoon et al. [6] (compressive strength of 238 MPa and bending strength of 94 MPa) due to porosity.

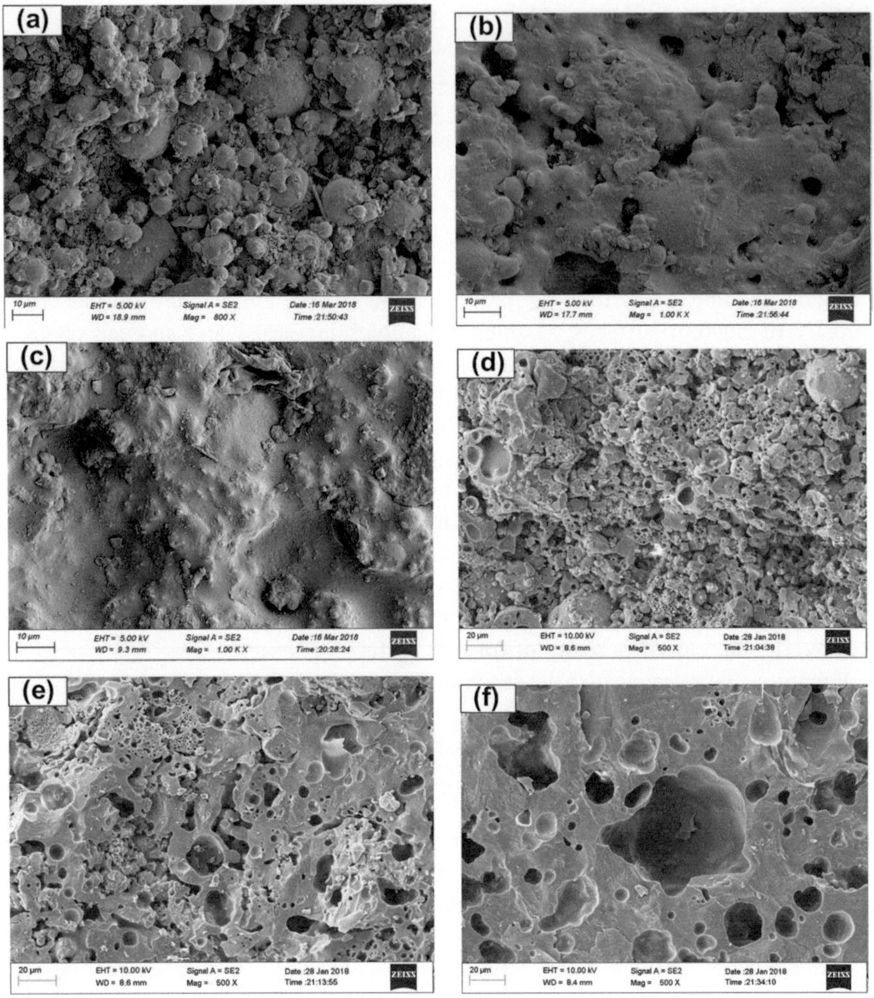

Fig. 6 SEM micrographs of sintered samples: **a** CFA1000-60 (surface), **b** CFA1100-60 (surface), **c** CFA1180-30 (surface), **d** CFA1000-60 (fracture surface), **e** CFA1100-60 (fracture surface), and **f** CFA1180-30 (fracture surface)

The strength between the porous material (δ) and the same material without porosity (δ_0) can be calculated by the formula proposed by Rice: $\frac{\sigma}{\sigma_0} = (1 - AP)^n$ [24]. The porosity of CFA1180-30 was estimated to 26% (P = 0.26), with n = 3 and A = 1 (for spherical pores) [17]. The compressive and bending strength of glass-ceramics produced by direct sintering should be 98 MPa and 38 MPa, respectively. The measured compressive strength was hence larger than the estimated, while the measured bending strength was lower. This could be due to different locations of remarkable stress concentrations.

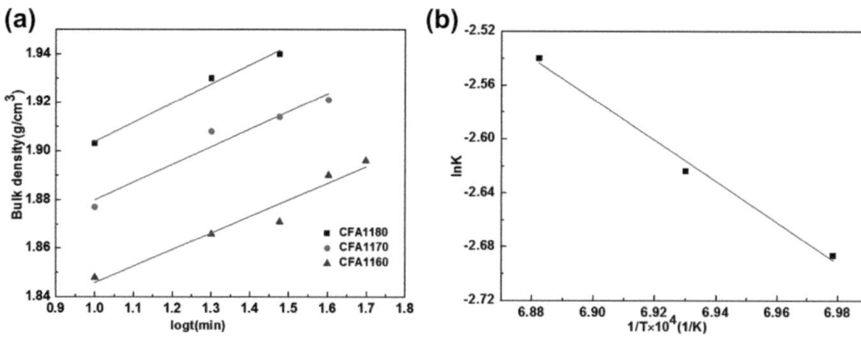

Fig. 7 **a** Bulk density (D) versus log*t* graph and **b** *ln* k versus 1/T graph

Table 2 The physical and mechanical properties of CFA1180-30

Sample	Sintering temperature (°C)	Holding time (min)	Bulk density (g cm^{-3})	Water absorption (%)	Compressive strength (MP)	Bending strength (MP)
CFA1180-30	1180	30	1.94	1.04	122	34

Conclusions

Glass-ceramics were prepared from coal fly ash by direct sintering to solve the environmental problems and offering large-scale manufacturing. The XRD and SEM analyses of sintered samples indicated that increased sintering temperature from 1000 to 1180 °C dissolved the amorphous metastable glassy phase, hematite, and portion quartz to form anorthite during sintering densification process. The sintering activation energy of glass-ceramics from CFA was estimated to 127 kJ/mol indicating CFA could easily be sintered not only due to high special surface energy of CFA powder but also to the short sintering time. The water absorption, bulk density, compressive and bending strength of samples sintered at 1180 °C for 30 min were estimated to 1.0%, 1.94 g cm^{-3}, 122 MPa, and 34 MPa, respectively. These values appeared promising for potential applications in lightweight construction tiles, such as coverage of internal wall and manufacturing of ventilated facades. Overall, this study did not only investigate how to maximize used fly ash to provide industrial production method of glass-ceramics but also established comprehensive utilization of similar solid wastes containing amorphous glassy phase and valuable oxide resources of glass-ceramics.

Acknowledgements This work was financially supported by the Longshan academic talent research and Innovation Team Project of SWUST (Grant No. 17LZXT11).

References

1. Qi L, Yuan Y (2011) Characteristics and the behavior in electrostatic precipitators of high-alumina coal fly ash from the Jungar power plant, Inner Mongolia, China. J Hazard Mater 192:222–225
2. Liu J, Dong Y, Dong X et al (2016) Feasible recycling of industrial waste coal fly ash for preparation of anorthite-cordierite based porous ceramic membrane supports with addition of dolomite. J Eur Ceram Soc 36:1059–1071
3. Luo Y, Ma S, Liu C et al (2017) Effect of particle size and alkali activation on coal fly ash and their role in sintered ceramic tiles. J Eur Ceram Soc 37:1847–1856
4. Kim JM, Kim HS (2004) Processing and properties of a glass-ceramic from coal fly ash from a thermal power plant through an economic process. J Eur Ceram Soc 24:2825–2833
5. Li B, Deng L, Zhang X et al (2013) Structure and performance of glass-ceramics obtained by bayan obo tailing and fly ash. J Non-Cryst Solids 380:103–108
6. Yoon SD, Lee JU, Lee JH et al (2013) Characterization of wollastonite glass-ceramics made from waste glass and coal fly ash. J Mater Sci Technol 29(2):149–153
7. Erol M, Küçükbayrak S, Ersou-Meriçboyu A (2007) Production of glass-ceramics obtained from industrial wastes by means of controlled nucleation and crystallization. Chem Eng J 132:335–343
8. Francis AA, Rawlings RD, Sweeney R et al (2002) Processing of coal ash into glass ceramic products by powder technology and sintering. Glass Technol 42(2):58–62
9. Dimech C, Cheeseman CR, Cook S et al (2008) Production of sintered materials from air pollution control residues from waste incineration. J Mater Sci 43:4143–4151
10. Binhussain MA, Marangoni M, Bernardo E et al (2014) Sintered and glazed glass-ceramics from natural and waste raw materials. Ceram Int 40:3543–3551
11. Dai S, Zhao L, Peng S et al (2010) Abundances and distribution of minerals and elements in high-alumina coal fly ash from the Jungar powder plant, Inner Mongolia China. Int J Coal Geol 81:320–332
12. Lu J, Lu Z, Peng C et al (2014) Influence of particle size on sinterability, crystallisation kinetics and flexural strength of wollastonite glass-ceramics from waste glass and fly ash. Mater Chem Phys 148:449–456
13. Lu Z, Lu J, Li X et al (2016) Effect of MgO addition on sinterability, crystallization kinetics, and flexural strength of glass–ceramics from waste materials. Ceram Int 42:3452–3459
14. Yürüyen S, Toplan HÖ (2009) The sintering kinetics of porcelain bodies made from waste glass and fly ash. Ceram Int 35:2427–2433
15. Demirkiran AS, Artir R, Avci E (2008) Effect of natural zeolite addition on sintering kinetics of porcelain bodies. J Mater Process Tech 203:465–470
16. Andreola F, Barbieri L, Karamanova E et al (2008) Recycling of CRT panel glass as fluxing agent in the porcelain stoneware tile production. Ceram Int 34:1289–1295
17. Bernardo E, Lazzari MD, Colombo P et al (2010) Lightweight porcelain stoneware by engineered CeO_2 addition. Adv Eng Mater 12:65–70
18. Lin B, Li S, Hou X et al (2015) Preparation of high performance mullite ceramics from high-aluminum fly ash by an effective method. J Alloy Compd 623:359–361
19. Wei Z, Hou J, Zhu Z (2016) High-aluminum fly ash recycling for fabrication of cost-effective ceramic membrane supports. J Alloy Compd 683:474–480
20. Ilic M, Cheeseman C, Sollars C, et al (2003) Mineralogy and microstructure of sintered lignite coal fly ash. Fuel 82:331–336
21. Zhang Z, Zhang L, Li A (2015) Development of a sintering process for recycling oil shale fly ash and municipal solid waste incineration bottom ash into glass ceramic. Waste Manage 38:185–193

22. Mukhopadhyay TK, Ghosh S, Ghosh J et al (2010) Effect of fly ash on the physico-chemical and mechanical properties of a porcelain composition. Ceram Int 36:1055–1062
23. Qin J, Cui C, Cui X et al (2015) Recycling of lime mud and fly ash for fabrication of anorthite ceramic at low sintering temperature. Ceram Int 41:5648–5655
24. Rice RW (1998) Porosity of ceramics. Marcel Decker Inc., New York

Porous Cordierite Prepared by Emulsion Template Method Based on the First-Principles Calculation

Xuezhu Luan, Kaiyue Huang, Siyi Zhao and Jinhong Li

Abstract Porous cordierite ceramics (PCCs) were prepared by using emulsion template method using pure oxide as raw materials. The cell microstructures were constructed by the first-principle calculation. The thermal stability of cordierite was verified by the lattice vibration calculation at high temperature, band gap, and DOS. The microstructures and the effects of solid content on mechanical properties were investigated. PCCs with solid content of 40% had the uniform pore structure, which can be used in special industries.

Keywords Cordierite · First-principle calculation · Thermal stability · Porous structure · Solid content

Introduction

With the continuous improvement control of porous materials, the application fields and application scope of porous ceramics are also expanding [1, 2]. At present, its applications have spread to many scientific fields, such as environmental protection, energy saving, chemical industry, petroleum, smelting, food, pharmaceuticals, biomedicine, and so on [3–6]. The scientific fields of smelting, food, pharmaceuticals, and biomedicine have attracted the attention of global materials science [3–10]. Among them, porous cordierite ceramics (PCCs) can be used under relatively severe working conditions due to their extremely low thermal expansion coefficient and excellent thermal shock resistance, and become a common catalyst carrier material and high temperature and high pressure filter [11–16]. Currently, it is mainly made by artificial synthesis in industry. Compared with naturally occurring cordierite, synthetic cordierite has superior performance. With the

X. Luan · K. Huang · S. Zhao · J. Li (✉)
Beijing Key Laboratory of Materials Utilization of Nonmetallic Minerals and Solid Wastes, National Laboratory of Mineral Materials, School of Materials Science and Technology, China University of Geosciences, Beijing 100083, People's Republic of China
e-mail: jinhong@cugb.edu.cn

B. Li et al. (eds.), *Advances in Powder and Ceramic Materials Science*, The Minerals, Metals & Materials Series,
https://doi.org/10.1007/978-3-030-36552-3_2

continuous efforts of scientists for many years, porous cordierite ceramics have been developed to a great extent, and have achieved great economic and social benefits, but there are still several aspects to be further studied and improved [17–20]. Thermomechanical properties, the relationship between porosity and strength, precise control methods for pore size and shape distribution of materials, and reduction of production costs.

In this experiment, PCCs were prepared by using emulsion template method using pure oxide as raw materials. The microstructures and thermal stability were determined by first-principles calculation. The effects of solid content on mechanical properties were investigated.

Experiments

Commercially available materials, magnesia (AR, Sinopharm Chemical Reagent Co., Ltd. Shanghai, China), alumina (AR (analytical reagent), Xuancheng Jingrui New Material Co., Ltd. Anhui Province, China), and silica (AR, Shenzhen Crystal Materials Chemical Co., Ltd. Guangdong Province, China), were blended together to obtain ceramic suspensions and were ball milled for 12 h with ZrO_2 balls (10 mm), 2.0 wt% ammonium polyacrylate salt (Adamas Reagent Co., Ltd.), 0.5 wt% PVA (AR, Adamas Reagent Co., Ltd.) and different addition of mullite fibers (Beijing Chemical Works). Afterward, 2.0 wt% propyl gallate ($C_{10}H_{12}O_5$) (AR, Adamas Reagent Co., Ltd.) was dissolved in ethanol (AR, Beijing Chemical Works) and added to suspended particles, until arriving an in situ modified state [21]. Then, octane (200 vol%, AR, Beijing Chemical Works) was added to the suspension, followed by the emulsification process under continuous stirring by an electric mixer (HR1613, Philips) at full speed for 5 min. Finally, all the emulsions were transferred to culture vessels (2 cm in height and 9 cm in diameter) and dried at room temperature. After oil and water evaporate, the dried samples were shaped. Finally, the samples were sintered at different temperatures for 2 h. The morphology was observed by scanning electron microscopy (SEM, S4800, Hitachi, Japan). The pore size distribution of the samples was evaluated by mercury intrusion porosimeter (PoreMasterGT60, Quantachrome).

Results and Discussion

The energy band theory is a major theoretical basis for studying microscopic motion. It is based on the viewpoint of quantum mechanics, so that individual electrons in a solid are no longer bound to individual atoms, but move throughout the whole solid. Although the single electron theory is an approximation theory, its actual development proves that it has become the basis for accurately summarizing the laws of electron motion in some important fields. The band theory divides crystals into metals, semiconductors, and insulators according to electrical properties. It not only explains the conductive properties of the crystal but also explains the optical, magnetic. and thermal properties of the crystal. The energy band structure calculated by the first principle can explain the difference between metal, semiconductor, and insulator. When the band gap is 0 eV or very small, the solid is a metal material, and electrons can easily obtain energy at room temperature and transition to rewinding to conduct electricity. When the band gap is greater than 9 eV, it is an insulating material, and it is difficult for electrons to transition to the conduction band, and thus it is impossible to conduct electricity. The semiconductor material has a band gap of 1–3 eV between the conductor and the insulator. As shown in Fig. 1, the cordierite microstructure was closely combined by SiO_2, Al_2O_3 tetrahedron and MgO octahedron, and the gap of the band gap was 4.6875 eV. The band length was a transition, so cordierite had a stable structure. As shown in Fig. 2, it can be seen from the state density analysis that the DOS state had the X-axis symmetry characteristic and belong to the semiconductor or insulator, which further verified the thermal stability of cordierite. It can be seen from Fig. 3 that the microstructure of cordierite was calculated at high temperature, and its microstructure did not change at room temperature and high temperature of 1300 °C, so cordierite had excellent high-temperature resistance.

The effect of different solid contents on the morphology of PCCs is shown in Fig. 4. When the solid content was 20%, the particles and particles cannot be bridged together because of fewer solid particles. When the oil phase was removed, the spherical structure collapsed to form an irregular macroporous structure. With the increase of solid content, the pore structure tended to be uniform. When the solid content was 40%, the porous structure with uniform pore size distribution was formed, and PCCs were formed by high temperature sintering. As shown in Fig. 5, the pore wall became thinner after high temperature sintering. Cordierite, as rod-like particles, was arranged on the pore wall after high temperature sintering to

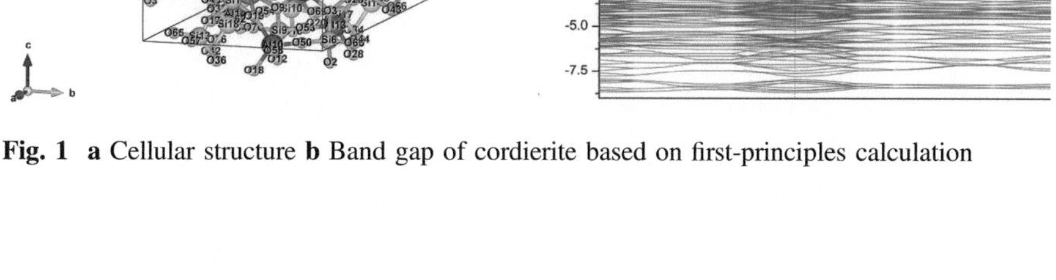

Fig. 1 **a** Cellular structure **b** Band gap of cordierite based on first-principles calculation

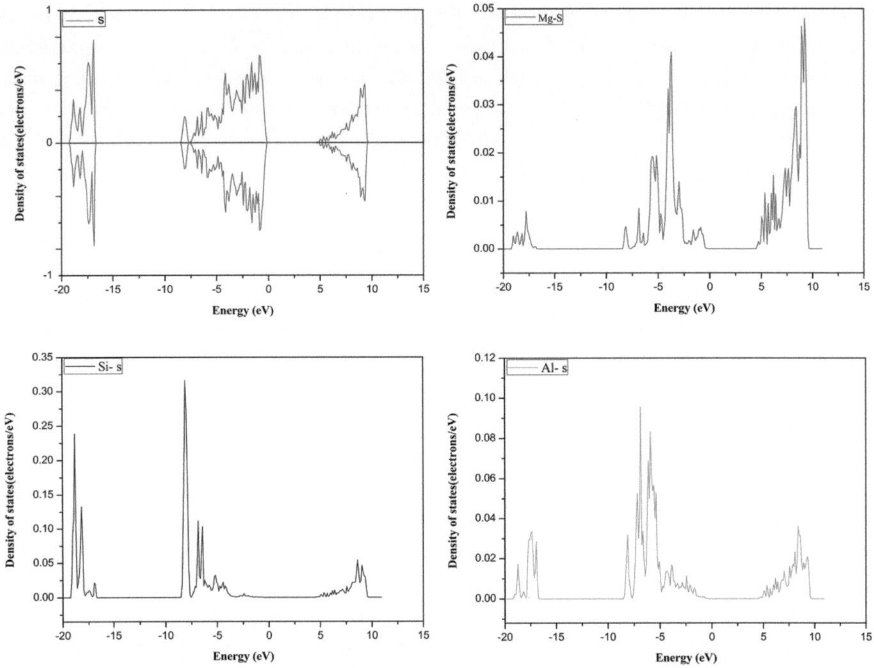

Fig. 2 DOS of cordierite

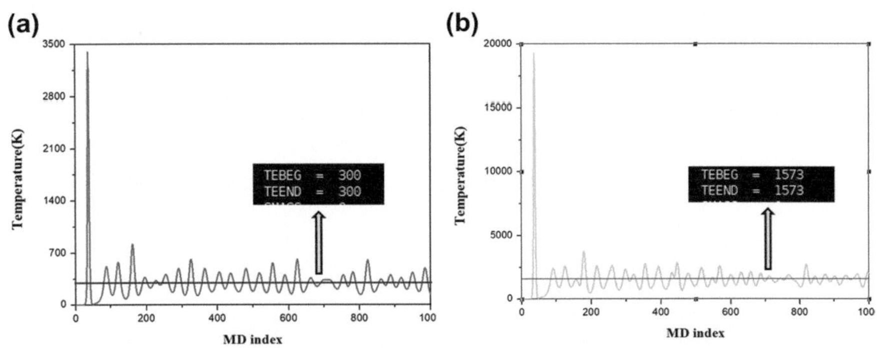

Fig. 3 First-principles calculation of cordierite at high temperature

form porous structure. The strength of PCCs is mainly determined by the properties of the products. The second impurity phase will not appear by pure oxide as raw materials. Cordierite as the main crystal phase can improve its thermal stability and the mechanical strength.

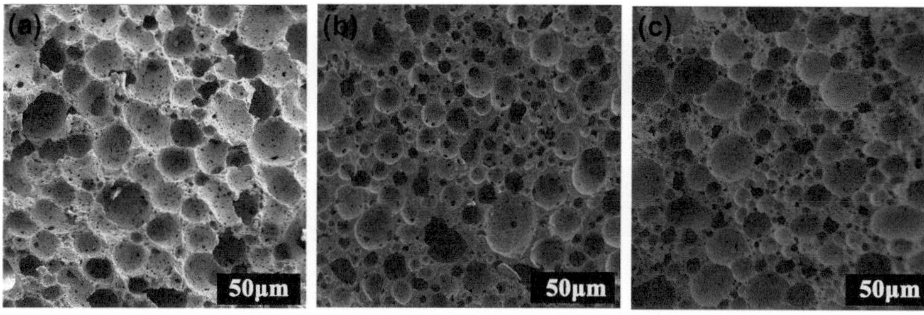

Fig. 4 The effect of different solid contents on the morphology of PCCs **a** 20%, **b** 30%, **c** 40%

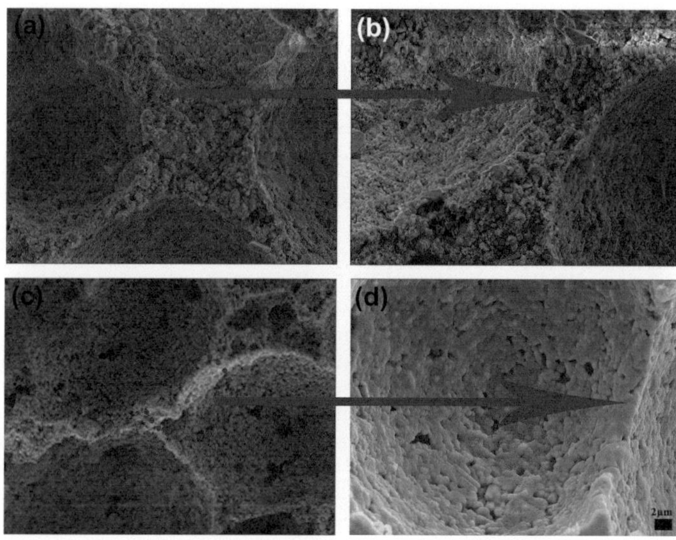

Fig. 5 SEM images of PCCs at different temperatures **a** 0 °C, **b** pore wall at 0 °C, **c** 1300 °C, **d** pore wall at 1300 °C

Conclusions

In this experiment, PCCs were prepared by using an emulsion template method using pure oxide as raw materials. The first-principle calculation was used to construct the cell microstructures to verify its excellent properties. Its stability was verified by band gap. DOS verified its non-conductivity. The thermal stability was verified by the lattice vibration calculation at high temperature. The microstructures and the effects of solid content on mechanical properties were investigated. PCCs with solid content of 40% had the uniform pore structure, which can be used in special industries.

Acknowledgements This work was supported by the National Natural Science Foundation of China (Grant No. U1607113 and U1803116), Program of Qinghai Science and Technology Department (Grant No. 2017-HZ-8058).

References

1. Wang S, Ma X-Y, Wang Y-L, Cui S-P, Nie Z-R, Li Q-Y, Wei Q (2019) Preparation and desalination performance of porous planar cordierite membranes using industrial solid waste as main silica source. Ceram Int 45:5932–5940
2. Hu Y, Xiao Z, Wang HP, Ye C, Wu YQ, Xu SQ (2019) Fabrication and characterization of porous $CaSiO_3$ ceramics. Ceram Int 45:3710–3714
3. Deng X, Wu Y, Wei T, Ran S, Huang L, Zhang H, Li F, Han L, Ge S, Zhang S (2018) Preparation of elongated mullite self-reinforced porous ceramics. Ceram Int 44:7500–7508
4. Gao HT, Liu XH, Chen JQ, Qi JL, Wang YB, Ai ZR (2018) Preparation of glass-ceramics with low density and high strength using blast furnace slag, glass fiber and water glass. Ceram Int 44:6044–6053
5. Qiu L, Zou H, Tang D, Wen D, Feng Y, Zhang X (2018) Inhomogeneity in pore size appreciably lowering thermal conductivity for porous thermal insulators. Appl Therm Eng 130:1004–1011
6. Sandoval ML, Talou MH, Tomba Martinez AG, Camerucci MA, Gregorová E, Pabst W (2018) Porous cordierite-based ceramics processed by starch consolidation casting–microstructure and high-temperature mechanical behavior. Ceram Int 44:3893–3903
7. Li Y, Li J, Feng W, Wang X, Nian H (2017) Design and preparation of the phase change materials paraffin/porous Al_2O_3@graphite foams with enhanced heat storage capacity and thermal conductivity. ACS Sustain Chem Eng 5:7594–7603
8. Naga SM, Sayed M, Elmaghraby HF, Khalil MS, El-Sayed MA (2017) Fabrication and properties of cordierite/anorthite composites. Ceram Int 43:6024–6028
9. Shin DS, Kim HG, Ahn HS, Jeong HY, Kim Y-J, Odkhuu D, Tsogbadrakh N, Lee H-B-R, Kim BH (2017) Distribution of oxygen functional groups of graphene oxide obtained from low-temperature atomic layer deposition of titanium oxide. RSC Adv 7:13979–13984
10. Li J, Ma Y, Wang X, Liu L, Wang C (2017) Highly permeable macroporous cordierite ceramics with controlled microstructure produced by particle-stabilized emulsions with a reactive thermal treatment. J Eur Ceram Soc 37:3203–3211
11. Pia G, Casnedi L, Sanna U (2016) Porosity and pore size distribution influence on thermal conductivity of yttria-stabilized zirconia: experimental findings and model predictions. Ceram Int 42:5802–5809
12. Tokizono T, Tsuru Y, Atsumi T, Hosokawa N, Ohnuma T (2016) Theoretical approaches for studying anisotropic negative thermal expansion: a case of cordierite. J Ceram Soc Jpn 124:744–749
13. Wang F, Yin JW, Yao DX, Xia YF, Zuo KH, Xu JQ, Zeng YP (2016) Fabrication of porous SiC ceramics through a modified gelcasting and solid state sintering. Mater Sci Eng A Struct Mater Prop Microstruct Process 654:292–297
14. Wang X, Li J-H, Xie Y-M, Zhang H-Y (2016) Three-dimensional fully interconnected highly porous hydroxyapatite scaffolds derived from particle-stabilized emulsions. Ceram Int 42:5455–5460
15. Zhang RB, Qu Q, Han BY, Wang BL (2016) A novel silica aerogel/porous Y_2SiO_5 ceramics with low thermal conductivity and enhanced mechanical properties prepared by freeze casting and impregnation. Mater Lett 175:219–222
16. Baitalik S, Kayal N (2017) Processing and properties of cordierite-silica bonded porous SiC ceramics. Ceram Int 43:14683–14692

17. Liu T, Tang Y, Li Z, Wu T, Lu A (2016) Red mud and fly ash incorporation for lightweight foamed ceramics using lead-zinc mine tailings as foaming agent. Mater Lett 183:362–364
18. Lu J, Cong X, Lu Z (2016) Influence of magnesia on sinter-crystallization, phase composition and flexural strength of sintered glass-ceramics from waste materials. Mater Chem Phys 174:143–149
19. Lu Z, Lu J, Li X, Shao G (2016) Effect of MgO addition on sinterability, crystallization kinetics, and flexural strength of glass–ceramics from waste materials. Ceram Int 42:3452–3459
20. Ma N, Du LJ, Liu WT, Zhang XY, Huo WL, Qu YN, Gan K, Wang YL, Yang JL (2016) Preparation of porous Si_3N_4 ceramics with unidirectionally aligned channels. Ceram Int 42:9145–9151
21. Gonzenbach ARSUT, Tervoort E, Gauckler LJ (2006) Ultrastable particle-stabilized foams. Angew Chem 118:3606–3610

Preparation of Abrasion and Erosion-Resistant Ceramic Coating on Copper by Slurry Method

Zefei Zhang, Hao Bai, Ning Li, Jian Zhang and Huanmei Yuan

Abstract Copper is usually applied in severe environments such as extreme temperature, wear, or corrosive conditions which seriously affect the service life of copper equipment. In this study, a ceramic coating was prepared by slurry method. The slurry which was coated on the bond layer of NiCoCrAlY was blended by aggregates of SiO_2, Al_2O_3, and ZrO_2 and binder of sodium silicate. The ceramic top coating was formed through sintering at 600 °C. The results show that the $ZrSiO_4$ and $Al_2Si_3O_9$ which play a key role in improving the properties are formed as the main phases in the ceramic coating. Furthermore, the bonding strength was tested to be 10.29 MPa and the wear rate reached 3.040×10^{-5} $mm^3/(m \cdot N)$. After the test of 50 thermal cycles, the ceramic coating kept good condition. Thus, the coating prepared shows potential for copper protection in harsh environments.

Keywords Ceramic coating · Slurry method · Copper · Abrasion resistant Erosion resistant

Introduction

Copper and its alloys are widely used in electronic industries, marine industries, power stations, heat exchangers, and cooling tower industries [1] due to some favorable properties such as high electrical and thermal conductivity, mechanical workability, and malleability [2]. However, severe environments such as extreme temperature, wear, or corrosive conditions seriously affect the service life of copper equipment. And due to defects in wear, corrosion, and heat resistance the

Z. Zhang · H. Bai (✉) · N. Li · J. Zhang · H. Yuan
State Key Laboratory of Advanced Metallurgy, University of Science and Technology Beijing, 30# Xueyuan Road, Beijing 100083, China
e-mail: baihao@metall.ustb.edu.cn

School of Metallurgical and Ecological Engineering, University of Science and Technology Beijing, 30# Xueyuan Road, Beijing 100083, China

B. Li et al. (eds.), *Advances in Powder and Ceramic Materials Science*,
The Minerals, Metals & Materials Series,
https://doi.org/10.1007/978-3-030-36552-3_3

application range of copper is limited. Hence, it is necessary to improve its performance and service life.

An economic and effective way to solve the problem is to use protective ceramic coatings. Ceramic coating on metal surface can be prepared to obtain composite material, which has both the strength and toughness of the metal and the high-temperature resistance, abrasion resistance, corrosion resistance, and other advantages of ceramics [3, 4]. The technology of ceramic coating on metal surface has become a rapid developing direction of material development and application [5]. There are many methods of preparing metal-based ceramic coatings, such as laser cladding, physical or chemical vapor deposition (CVD), and thermal spraying. For example, Çelik [6] prepared Al_2O_3–TiO_2 ceramic coating on AZ31 magnesium alloy by plasma spraying method. Boudi et al. [7] investigated high-velocity oxy-fuel (HVOF) coating of Inconel 625 onto stainless steel and mild steel surfaces. But these methods are related to the complex process and their costs are high.

Slurry method (thermochemical reaction method) has attracted more and more attention in recent years [8]. The method has the advantages of easy process, no special equipment, convenient operation, and low cost, which is more suitable for industrial applications. Slurry, suspension of ceramic aggregates in a binder–solvent system, can be uniformly coated on complex substrate by means of a brush, immersion, or spraying. After that, the coated specimen is subjected to a curing heat treatment. During the curing and heating, thermochemical reactions will take place between the aggregates, adhesives and the matrix, and some composite ceramic phases will be formed, so as to improve the performance. Considering this aim, it is desirable to develop a ceramic coating by slurry method, to protect copper in harsh environments.

In the present work, we prepared ceramic coatings on copper substrate, by using SiO_2, Al_2O_3, and ZrO_2 as ceramic aggregates and sodium silicate as binder. The phase composition and microstructure of ceramic coatings were analyzed. And the wear resistance and corrosion resistance of ceramic coatings were tested.

Experimental Procedure

Coating Preparation

The main raw materials used as ceramic aggregates were SiO_2 (analytical reagent, China), Al_2O_3 (Analytical Reagent, China), and ZrO_2 (analytical reagent, China). Sodium silicate water glass (chemically pure, China) was used as the coating binder and the molar ratio of SiO_2/Na_2O is 3.3. The slurry preparation proceeded as follows: The different ceramic aggregates were mixed in the proportions reported in Table 1. Then, the mixed ceramic aggregates were added to water glass solution with a 35%:65% mass proportion to obtain mixed ceramic slurry.

Table 1 Formula of ceramic slurry system

Chemical compositions	SiO_2	Al_2O_3	ZrO_2
Content/%	47	29	24
Granularity/μm	40	10	2

Table 2 Parameters of plasma spraying

Parameter	Values	Parameter	Values
Flow rate of primary gas	45 L/min	Spray distance	100 mm
Feedstock giving rate	40 g/min	Power	500 A × 70 V
primary gas	Ar	Secondary gas	H_2
Gun moving rate	800 mm/s	Spray inclinations	90°

Matrix material was copper sheet. The copper was cut into specimen plates with dimensions Φ25 mm × 5 mm for the phase and microstructure characterization of the coating and 20 mm × 20 mm × 5 mm for the performance test of coating using a wire cutting machine. All of these specimens were cleaned in acetone for 30 min and in an ultrasonic bath containing an alcohol solution for 6 h, to remove the grease on the surface of the matrix. The cleaned copper specimens were sandblasted with 60 mesh sand in a box type sandblasting machine, for two purposes. One is to remove the oxide scale and the other is to increase the surface roughness and strengthen the bond between the substrate and the coating.

Firstly, The NiCoCrAlY bond coating with a thickness of about 100 μm was plasma-sprayed (Sulzer Metco 9MC, America) on the copper surface. The spraying parameters are shown in Table 2. Next, the homogeneous ceramic slurry was brushed to coat onto the specimen and before coated, the surface of specimens was cleaned by alcohol. Subsequently, the coated specimen dried in shade for 5 h, and then put it in drying oven (YDYQ XGQ-3000, China) at 85 °C for 5 h. Finally, the coated specimen was sintered at 600 °C for 4 h in a muffle furnace (KSL-1100X-L, China).

Coating Characterization

Phase composition analysis of coating was carried out by X-ray diffraction (XRD, Smart Lab, Japan). The cross section was first polished with coarse sandpaper, then with metallographic sandpaper 0#, 1#, 2#, 3#, 4#, and 5# until the surface scratches were small and uniform, and then polished with diamond polishing gypsum, thrown to the cross section bright without scratches. The surface and cross section microstructure of the coating was observed by scanning electron microscopy (SEM, ZEISS EVO18, Germany).

Testing of Thermal Shock Resistance

The thermal shock resistance of the coating was evaluated by water quenching method. The samples were put into a muffle furnace (KSL-1100X-L, China), heated to 400 °C and preserved for 10 min, then taken out and quickly quenched into water at room temperature. After drying, the coating was observed. If there is no crack or spalling on the surface, it will be regarded as a thermal shock cycle. Repeat this process until nearly one-third of the coating peeled off. The average value was calculated as a criterion for evaluating the thermal shock resistance of the coatings.

Bonding Strength Test

The bond strength of coating was measured by dual sample tensile method. E-7 high temperature adhesive was selected as the interface binder in this test. The back of the specimen with coatings was coarsened by sandpaper. After that the specimen was bonded dual tension rods by adhesive, and then clamped with fixtures, whose contact pressure was 0.05 MPa. Tensile specimens are first placed at room temperature for 3 h, then heated to 105 °C and cured for 3 h in a furnace, finally cooled in air for tensile testing. In this experiment, a computer controlled electronic universal testing CMT5105 machine (SANS Testing Machine Co., Ltd., Shenzhen, China) was used at a loading speed of 1 mm/min. Three samples were tested, and then the average value was taken as the measured bond strength of the coating.

Wear and Corrosion Test

The friction and wear properties of the coatings were measured by reciprocating friction and wear tester (MFW-02, China). The friction pair was Si_3N_4 ceramic ball with diameter of 6.35 mm. The load was 600 g (5.88 N), the reciprocating stroke was 10 mm, the reciprocating frequency was 3 Hz, and the wear time was 55.6 min. The number of tests was 10,000 times, and the limit value of the wear mark depth was 100 um. The volume wear rate of the coatings was measured by a LEXT OLS4100 laser scanning confocal microscope (Olympus Corporation, Japan). The wear rate v_d ($mm^3/(m \cdot N)$) is calculated as follows:

$$v_d = S \times L/F \times l, \tag{1}$$

where S(mm^2) is the cross-sectional area of wear scar, L(mm) is the length of wear scar, F(N) is load, and l(m) is wear distance.

The corrosion resistance of ceramic coating is an important criterion to qualify the coatings for the desired applications. Since corrosive gases like chlorine and sulfur usually cause severe damage to the copper in service, the coating was subjected to corrosion testing by high-temperature corrosion method. In detail, $FeCl_3$ and $FeSO_4$ salts were mixed in a ratio of 50 and 50 wt%. After that, it was covered uniformly on the coating surface with coverage of 25 mg/cm^2 [9]. Subsequently, the sample was isothermally heated to 600 °C in a furnace for 15 h in air atmosphere and then cooled to room temperature, maintaining the heating and cooling rates at 4 °C/min.

Results and Discussion

Structure Analysis

Ceramic coating was prepared by coating ceramic slurry on the surface of the matrix, curing at room temperature after overcast and dry at high temperature. During high-temperature curing, thermochemical reaction will occur to produce new composite ceramic phases [10]. Figure 1 shows the XRD patterns of ceramic coating. From Fig. 1, the ceramic coating contains three raw material phases of SiO_2, Al_2O_3, and ZrO_2 indicating that the reactions took place only partially. The new phases produced in ceramic coating are $ZrSiO_4$ and $Al_2Si_3O_9$, which are due to the two reactions during high temperature sintering process.

$$SiO_2 + ZrO_2 \rightarrow ZrSiO_4 \tag{2}$$

$$3SiO_2 + Al_2O_3 \rightarrow Al_2Si_3O_9 \tag{3}$$

$ZrSiO_4$, the main component of zircon, has high hardness, high-temperature resistance, thermal stability and corrosion resistance [11]. $Al_2Si_3O_9$ is a

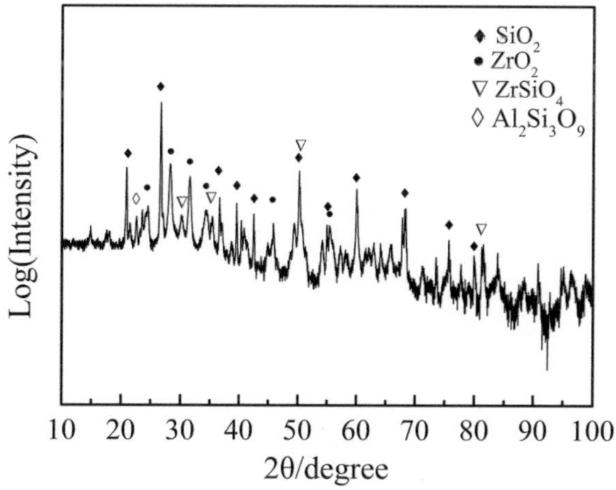

Fig. 1 XRD pattern of the ceramic coating

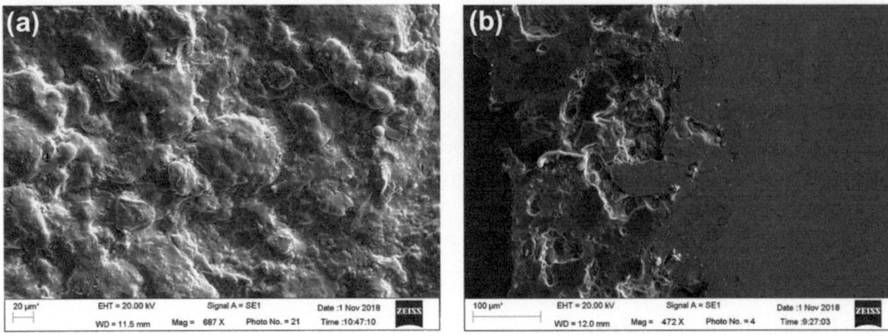

Fig. 2 The surface (**a**) and cross section (**b**) SEM micrographs of the ceramic coating

high-temperature resistant phase and has stable chemical and physical properties. The two new phases, as the main phases in the ceramic coating, play a key role in improving the properties.

Figure 2 shows the surface morphology and cross section morphology of the ceramic coating. Figure 2a shows that the surface of the ceramic coating is compact with no obvious defects. It can be seen from Fig. 2b that the ceramic top coating was uniform and tightly bonded with the matrix, and the bond coating was also tightly bonded with the copper matrix. Two reasons could explain the phenomenon. The one was that sodium silicate water glass is a film with a three-dimensional network structure and excellent water resistance, which can form a close bond with the ceramic aggregates and a relatively strong bond with the substrate. The other was some solid–solid state reactions 1 and 2 had occurred in the coating interior at 600 °C during the thermal curing for 4 h, the interface atoms had enough time for mutual diffusion.

Bonding Strength

The bonding strength of coatings is an important index of coatings performance. Good bonding between coatings and matrix is the basic condition for coatings to play their own functional role [12]. Table 3 shows the experimental data of bonding strength of ceramic coating. From Table 3, the average bonding strength can reach to 10.29 MPa. After curing at 600 °C, there were reactions between the ceramic coating and the bond coating, the matrix and the bond coating. Therefore, there is not only physical bonding but also chemical bonding between the coating and the matrix and the bond coating, which enhances the bonding strength between them.

Table 3 Bonding strength of ceramic coating

Samples	1	2	3	Average value
Content/kN	4.9	4.95	5.3	5.05
Section diameter/mm	25	25	25	25
Bonding strength/Gpa	9.98	10.09	10.80	10.29

Thermal Cycling Studies

Table 4 shows the results of thermal shock resistance test of coatings. It is generally considered that thermal stress is a key factor affecting the thermal shock resistance of coatings. So, it is very important to match the thermal physical properties (thermal expansion coefficient and elastic modulus) of coating and substrate [13]. From Table 4, the thermal cycles of ceramic coatings at 400 °C are higher than 50 times. In addition to the high bonding strength of the foundation, because the thermal expansion coefficient of NiCoCrAlY (about 15.3–15.5 $\times$ 10^{-6} K^{-1}) is between that of copper (about 18 $\times$ 10^{-6} K^{-1}) and ceramic top coating (about 10 $\times$ 10^{-6} K^{-1}). Using NiCoCrAlY as transition coating material, the abrupt change of physical properties and the difficult bonding between the coating and the copper substrate is eliminated, the bonding condition between the coatings is improved, and the preparation stress in the coating is alleviated.

Wear Resistance

Figure 3 shows the morphology of the wear track of the ceramic coating after dry sliding wear test. It can be seen from the figure that the wear surface of the coating is uneven and there are small pits. The wear track on the coating surface is very shallow and narrow showing good wear resistance. In order to accurately calculate wear rate, the cross-sectional profile of the wear track of the coating was obtained by a LEXT OLS4100 laser scanning confocal microscope (Olympus Corporation, Japan), as shown in Fig. 4. It can be concluded from the figure that the wear mark depth of the coating is 10 μm and the width is 750 μm. By calculating the integral area of the corresponding wear trace profile, the wear rate of the ceramic coating is calculated as 3.040 $\times$ 10^{-5} $mm^3/(m \cdot N)$, which is about four times that of copper matrix (1.238 $\times$ 10^{-4} $mm^3/(m \cdot N)$). The improvement of wear resistance is mainly related to the hardness and microstructure of the ceramic coating. The two new

Table 4 Thermal shock resistance of ceramic coating

Test temperature/°C	400
Thermal shock resistance results	50 cycles not falling off

phases evenly distribute in the ceramic coating, and the ceramic top coating is compact with no obvious defects.

High-Temperature Corrosion

Figure 5 shows the surface morphology of the uncoated and coated copper before and after 15 h of corrosion. During the high-temperature corrosion process, $FeCl_3$ and $FeSO_4$ salts were mixed uniformly and spread over the specimen's surface, and after that chlorine, sulfur dioxide, and iron oxide were produced according to Eq. 4, as corrosive media in many harsh environments.

$$FeCl_3(s) + FeSO_4(s) + 1/2O_2(g) \rightarrow Fe_2O_3(s) + 3/2Cl_2(g) + SO_2(g) \quad (4)$$

It can be seen from Fig. 5 that the copper has been seriously corroded, resulting in many irregular corrosion pits and uneven surface. And there is also gray-green corrosion product on the surface. However, the ceramic coating has no obvious cracks on the surface except for the brown iron oxide products formed by reaction Eq. 4. The coating effectively protects copper from high-temperature corrosion. The corrosion resistance of ceramic coating is improved compared to copper substrate, which is due to the fact that ceramic aggregates can improve the corrosion resistance of the coating to a certain extent. Not only that, through phase analysis, the new phases, reaction products during high temperature sintering process, have excellent corrosion resistance and improve the corrosion protection performance of the coating.

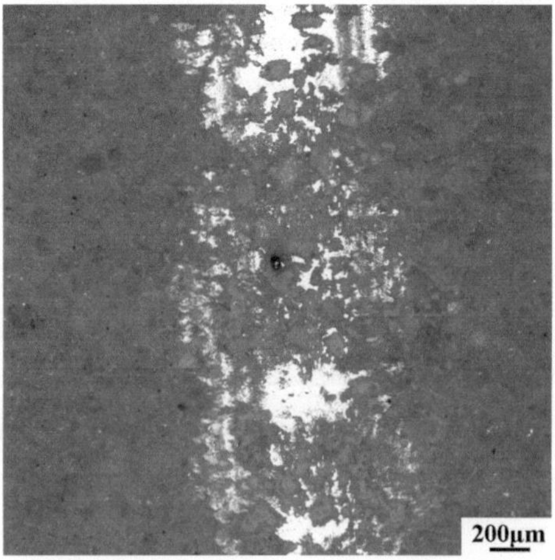

Fig. 3 The wear mark of the ceramic coating

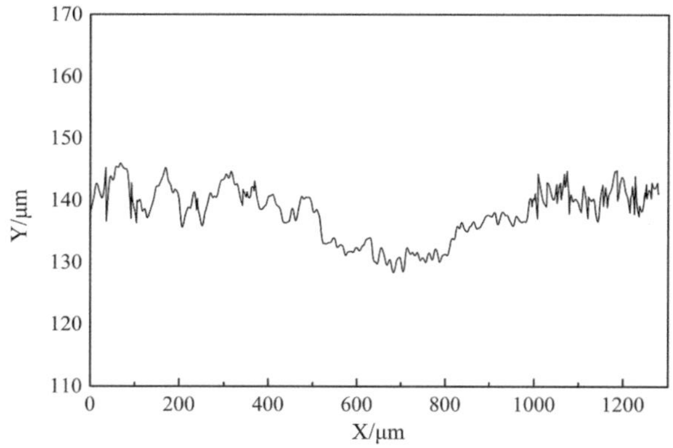

Fig. 4 The grinding line scan curve of ceramic coating

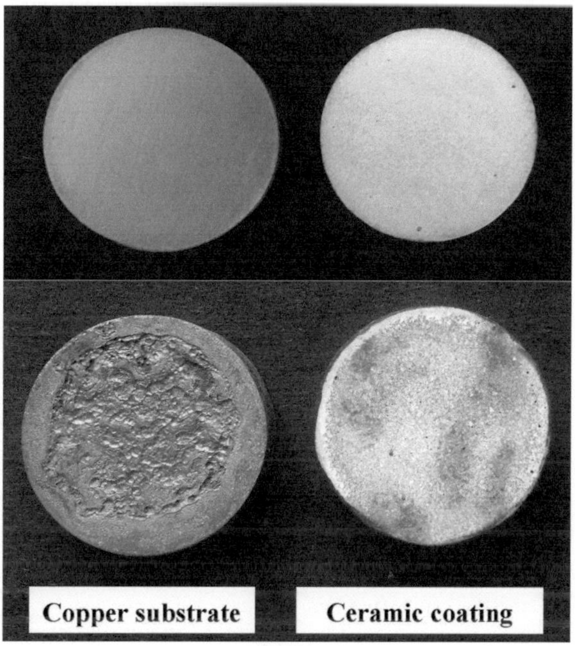

Fig. 5 The morphology before and after corrosion

Conclusions

(1) Ceramic coatings were prepared by slurry method using SiO_2, Al_2O_3, and ZrO_2 as aggregates and sodium silicate as adhesive. New phases $ZrSiO_4$ and $Al_2Si_3O_9$ were formed in the ceramic coatings, which enhanced the bonding strength between the coating and the substrate.

(2) NiCoCrAlY transition coating is helpful to the bonding strength and thermal shock resistance of the ceramic coating. The bonding strength can reach to 10.29 MPa. After the test of 50 thermal cycles, the ceramic coating kept good condition.
(3) The ceramic coating has good wear resistance and corrosion resistance owing to its good phase structure and structure so that copper can be protected from wear and corrosion.

Acknowledgements This work was supported by the Innovation Talents Fund Project of University of Science Technology Beijing, and the Shantou Science and Technology Project of Guangdong (No. 100).

References

1. Duran B, Bereket G, Duran M (2012) Electrochemical synthesis and characterization of poly (m-phenylenediamine) films on copper for corrosion protection. Prog Org Coat 73(2–3):162–168
2. Amin M, Khaled KF (2010) Copper corrosion inhibition in O_2-saturated H_2SO_4 solutions. Corros Sci 52(4):1194–1204
3. Luo H, Song P, Khan A, Feng J, Zang JJ, Xiong XP, Lü JG, Lu JS (2017) Alternant phase distribution and wear mechanical properties of an Al_2O_3-40 wt% TiO_2 composite coating. Ceram Int 43(9):7295–7304
4. Cao XQ, Vassen R, Tietz F, Stoever D (2006) New double-ceramic-layer thermal barrier coatings based on zirconia-rare earth composite oxides. J Eur Ceram Soc 26(3):247–251
5. Cheng B, Zhang YM, Yang N, Zhang M, Chen L, Yang GJ, Li CX, Li CJ (2017) Sintering-induced delamination of thermal barrier coatings by gradient thermal cyclic test. J Am Ceram Soc 100(5):1820–1830
6. Çelik İ (2016) Structure and surface properties of Al_2O_3–TiO_2 ceramic coated AZ31 magnesium alloy. Ceram Int 42(12):13659–13663
7. Boudi AA, Hashmi M, Yilbas BS (2004) HVOF coating of Inconel 625 onto stainless and carbon steel surfaces: corrosion and bond testing. J Mater Process Tech 155–156(none):2051–2055
8. Wang JQ, Yuan YC, Chi ZH, Zhang GX (2018) High-temperature sulfur corrosion behavior of h-BN-based ceramic coating prepared by slurry method. Mater Chem Phys 206:186–192
9. Fukubayashi HH (2006) Metal-zirconia composite coating. WO 2002/075004, 13 Dec 2006
10. Xiao K, Xue W, Li ZL, Wang JR, Li XM, Dong CF, Wu JS, Li XG, Wei D (2018) Effect of sintering temperature on the microstructure and performance of a ceramic coating obtained by the slurry method. Ceram Int 44(10):11180–11186
11. Cao XQ, Vassen R, Stoever D (2004) Ceramic materials for thermal barrier coatings. J Eur Ceram Soc 24(1):1–10
12. Wu HX, Ma Z, Liu L, Liu YB, Wang DY (2016) Thermal cycling behavior and bonding strength of single-ceramic-layer $Sm_2Zr_2O_7$ and double-ceramic-layer $Sm_2Zr_2O_7$/8YSZ thermal barrier coatings deposited by atmospheric plasma spraying. Ceram Int 42(11):12922–12927
13. An YL, Chen JM, Zhou HD, Liu GA (2010) Microstructure and thermal cycle resistance of plasma sprayed mullite coatings made from secondary mullitized natural andalusite powder. Surf Coat Tech 205(7):1897–1903

Part II
Advanced Ceramics and Processes

Toughening Mechanism of ZTA–TiC–Fe Ceramic Materials Produced by High-Gravity Combustion Synthesis

Hongwei Zhao, Shibin Guo, Jiangtao Li and Jinhong Li

Abstract Through coupling of high-gravity field and high temperature field of combustion synthesis, ZrO_2 toughened Al_2O_3 (ZTA) adding with TiC strengthened iron ductile (ZTA–TiC–Fe) composite materials were prepared by high-gravity combustion synthesis. The composite toughening of Al_2O_3–TiC ceramic materials was realized, including phase transformation toughening of tetragonal zirconia, zirconia fiber toughening, and metal ductile phase toughening. The results showed that the fracture toughness of ZTA–TiC–Fe composite materials reached 9–11 MPa·m$^{1/2}$, which was 2–3 times higher than pure Al_2O_3–TiC commercial tool materials.

Keywords Al_2O_3–TiC · Combustion synthesis · Zirconia toughened alumina

Introduction

Ceramic materials have the advantages of high strength, high hardness, and corrosion resistance, so they have a great prospect in industrial applications. However, due to their poor fracture toughness, their application scope is limited. Materials researchers have tried a variety of methods to improve the toughness of ceramic materials [1–3]. In recent years, the accepted solution is to conduct composite

H. Zhao (✉)
China Iron and Steel Research Institute Group, Beijing 100081, China
e-mail: zhw31032@126.com

H. Zhao · J. Li
School of Materials Science and Technology, China University of Geosciences, Beijing 100083, China

S. Guo · J. Li (✉)
State Key Laboratory of Technologies in Space Cryogenic Propellants, Technical Institute of Physics and Chemistry, Chinese Academy of Sciences, Beijing 100190, China
e-mail: lijiangtao@mail.ipc.ac.cn

B. Li et al. (eds.), *Advances in Powder and Ceramic Materials Science*, The Minerals, Metals & Materials Series,
https://doi.org/10.1007/978-3-030-36552-3_4

toughening of ceramic materials, which can greatly improve the toughness of ceramic materials [4–7].

Consolidated bulk ceramics are generally produced by powder sintering, such as pressureless sintering (PS) [8], hot pressing (HP) [9], and spark plasma sintering (SPS) [10]. The sintering method requires long-time heat treatment at high temperatures by furnace, which means a long processing time and much energy consumption. A recently-developed method to prepare bulk cermets is combustion synthesis under a high gas pressure or in a high-gravity field [11–13]. Combustion synthesis utilizes the heat energy released during exothermic combustion reactions to synthesize the target materials and does not require any furnaces [14, 15]. The high gas pressure or high-gravity field is applied to realize simultaneous densification during the synthesis for directly preparing dense bulk products. As a fast and furnace-free method, combustion synthesis exhibits a reduced processing time and near-zero energy consumption, which may be competitive in comparison with the conventional sintering approach.

In this paper, multi-component like ZrO_2 toughened Al_2O_3 (ZTA) adding with TiC strengthened iron ductile (ZTA–TiC–Fe) ceramic materials were prepared by high-gravity combustion synthesis. This material consists of zirconia toughened alumina ceramic matrix and antioxidant metallic iron reinforced by TiC hard particles. The toughening mechanism of the ZTA–TiC–Fe ceramic materials were investigated.

Experiment

Commercial powders of Al, Fe_2O_3, ZrO_2, Y_2O_3, Ti, and graphite were used as raw materials. The raw materials were mixed according to the following chemical reactions:

$$Al + Fe_2O_3 \rightarrow Al_2O_3 + Fe \tag{1}$$

$$3Al_2O_3 + 2ZrO_2 + 0.06Y_2O_3 = ZTA \tag{2}$$

$$Ti + C = TiC \tag{3}$$

The thermite system was used to produce Al_2O_3 and Fe, ZrO_2 and Y_2O_3 were used to produce the ZTA phase, and Ti and C were used to produce TiC. The molar ratio of Al_2O_3 and ZrO_2 in the ZTA phase was Al_2O_3:ZrO_2 = 3:2. In order to obtain high temperature of the reaction system, the weight ratio between ZTA and Fe was fixed to be ZTA:Fe = 3:2, and the weight content of TiC reinforcement phase was 20%.

The powder mixture of the raw materials was well homogenized by rotatory ball milling for 2 h with a rotation speed of 60 r/min, and then dried at 110 °C for 2 h. Each batch of 200 g reactant powder was cold-pressed into a round compact with a

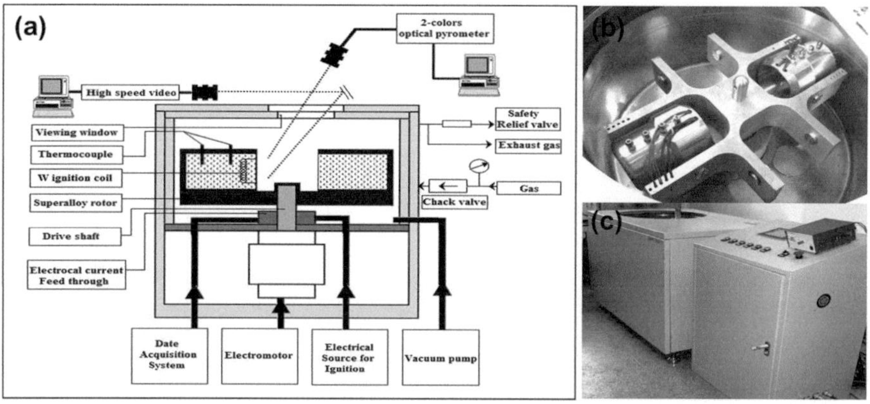

Fig. 1 A schematic diagram (**a**) and photos (**b**, **c**) of the facility for high-gravity combustion synthesis [11]

diameter of 40 mm at a pressure of 40 MPa, and the porosity in the compact was about 50%. The compact was placed in a graphite crucible. The crucible was coated with carbon felt and placed into a steel vessel, which was mounted on a rotator in a reaction chamber (see Fig. 1). A high-gravity field with an acceleration of 800 g (g = 9.8 m/s^2) was induced by the centrifugal effect. The combustion reaction was triggered by passing an electric current about 10 A through a tungsten coil closely above the reactant compact. When the reaction was over and the sample cooled down, the sample was taken out from the reaction chamber.

The synthesized samples were machined with an electroplated diamond wire saw and then well polished. The bulk density of samples was measured according to the Archimedes principle. The phase assemblage was identified by X-ray diffraction (XRD; D8 Focus, Bruker, Germany). The microstructure was examined by scanning electric microscopy (SEM; S4800, Hitachi, Japan). The Vickers hardness was measured by the indentation method with a load of 98 N and dwelling time of 15 s. The fracture toughness was tested by the single-edge-notched-beam (SENB) technique with bars of $25 \times 2 \times 4$ mm^3. The notch was generated with a diamond saw, and had a width of 0.3 mm and depth of 2 mm.

Results and Discussion

Figure 2 shows the XRD pattern of the synthesized ZTA–TiC–Fe ceramic material. Four crystalline phases of TiC, Al_2O_3, tetragonal ZrO_2, and Fe are observed agreeing well with the designated phase compositions. From the experimental results, the combustion reactions are complete and the reactants are fully converted into the target products.

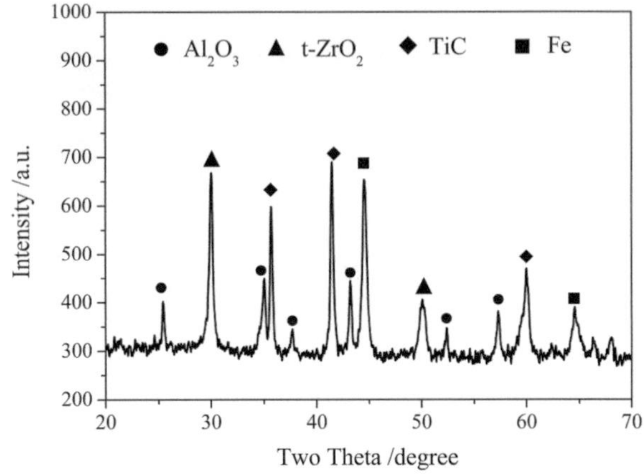

Fig. 2 XRD pattern of synthesized ZTA–TiC–Fe ceramic material

Figure 3 shows the microstructure features of the ZTA–TiC–Fe ceramic materials. The light Fe grains distribute uniformly in ZTA matrix. The TiC micro-grains locate at the interface between Fe grain and ZTA matrix. It forms a shell covering the surface of Fe particles, prevents the polymerization of Fe droplets, improves the wettability of metal Fe and ZTA ceramic matrix, and enhances the interfacial bonding strength of the two materials. Thus, this is conducive to improving the overall strength and toughness of the ZTA–TiC–Fe ceramic material.

The toughness of the ZTA–TiC–Fe ceramic material is 9–11 MPa·m$^{1/2}$, which is 2–3 times higher than pure Al_2O_3–TiC commercial tool materials. There are three toughening mechanisms that improved the toughness of the ZTA–TiC–Fe ceramic material. The first one is the ZrO_2 fibers toughened the Al_2O_3 matrix as show in Fig. 4a. The ZrO_2 fibers with 1–2 μm distributed uniformly in Al_2O_3 matrix, which has the same direction and equal interval. And that the ZrO_2 fibers are in situ

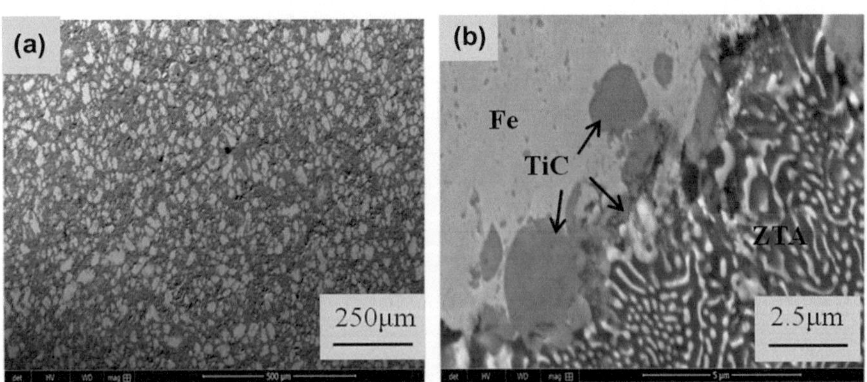

Fig. 3 Microstructure features of ZTA–TiC–Fe ceramic materials: **a** Fe uniform distribution in ZTA matrix, **b** TiC micro-grain at the interface between Fe and ZTA

formed in Al_2O_3 matrix with few defects at interface. It is better than the traditional method of extra addition. The second one is the phase transformation toughening of the tetragonal ZrO_2 as shown in Fig. 4b. There is obvious volume expansion at the fracture of the ZrO_2 fiber, this is due to tetragonal phase transformed monoclinic phase in ZrO_2 fiber by stress driving. The third one is ductile metal phase

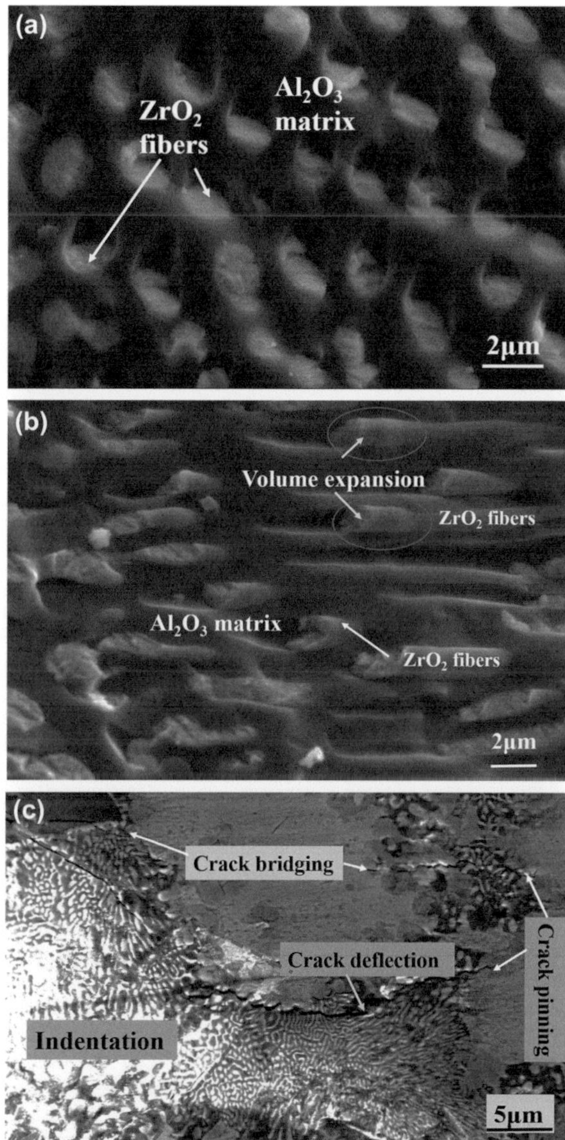

Fig. 4 Toughening mechanisms in ZTA–TiC–Fe ceramic material: **a** ZrO_2 fibers toughening; **b** ZrO_2 transformation toughening; **c** toughening by ductile metal phase

toughening as shown in Fig. 4c. The Fe grains cause the crack deflection for adding the crack propagation distance, and absorb some or whole stress of the crack tip appearing the crack bridging or crack pinning phenomenon.

Conclusion

ZTA–TiC–Fe ceramic material has been prepared by a fast and furnace-free way called high-gravity combustion synthesis. The Fe–TiC core-shell structure is conducive to distributing Fe grain in ZTA matrix and improving the interface strength. For the combination of three toughening mechanisms, such as fiber toughening, phase transformation toughening, and ductile phase toughening, the toughness of the ZTA–TiC–Fe ceramic material reaches 9–11 $MPa\cdot m^{1/2}$, which is 2–3 times higher than pure Al_2O_3–TiC commercial tool materials.

Acknowledgements This work is supported by the fund of the State Key Laboratory of Technologies in Space Cryogenic Propellants, SKLTSCP1802.

References

1. Becher PF (1991) Microstructural design of toughened ceramics. J Am Ceram Soc 74:256–269
2. Wang J, Stevens R (1989) Zirconia-toughened alumina (ZTA) ceramics. J Mater Sci 24:3421–3435
3. Kelly PM, Francis Rose LR (2002) The martensitic transformation in ceramics-its role in transformation toughening. Progress Mater Sci 47:463–557
4. Evans AG (1990) Perspective on the development of high-toughness ceramics. J Am Ceram Soc 73:187–206
5. Fang YH, Chen N, Du GP et al (2019) Effect of Y_2O_3-stabilized ZrO_2 whiskers on the microstructure, mechanical and wear resistance properties of Al_2O_3 based ceramic composites. Ceram Int 45:16504–16511
6. Arab A, Ibrahim Sktani ZD, Zhou Q et al (2019) Effect of MgO addition on the mechanical and dynamic properties of zirconia toughened alumina (ZTA) ceramics. Materials 12:2440–2453
7. Tan P, Wu P, Gao L et al (2019) Influence of Si_3N_4 content on the physical and mechanical properties of zirconia-toughened alumina (ZTA) ceramic composites. Mater Res Express 6:065205
8. Cai KF, McLachlan DS, Axen N et al (2002) Preparation, microstructures and properties of Al_2O_3–TiC composites. Ceram Int 28:217–222
9. Deng J, Can T, Sun J (2005) Microstructure and mechanical properties of hot-pressed Al_2O_3/TiC ceramic composites with the additions of solid lubricants. Ceram Int 31:249–256
10. Meir S, Kalabukhov S, Hayun S (2014) Low temperature spark plasma sintering of Al_2O_3–TiC composites. Ceram Int 40:12187–12192
11. Liu G, Li J, Yang Z, Guo S, Chen Y (2013) High-gravity combustion synthesis and in situ melt infiltration: a new method for preparing cemented carbides. Scripta Mater 69:642–645

12. Xi W, Li W, Xu T (2015) Thermite synthesis of TiC/FeNiCr cermet with double-layer structure. Mater Manu Proc 30:576–584
13. Liu GH, Li JT, Chen KX (2015) One-step preparation of dense $TiC_{1-x}N_x$–Ni_3Ti cermet by combustion synthesis. Mater Design 87:6–9
14. Merzhanov AG (2004) The chemistry of self-propagating high-temperature synthesis. J Mater Chem 14:1779–1786
15. Liu GH, Li JT, Chen KX (2013) Combustion synthesis of refractory and hard materials: a review. Int J Refract Met Hard Mater 39:90–102

Part III
Ceramic Nanoparticles and Powder

Pressure-Less Processing of Ceramics with Deliberate Elongated Grain Orientation and Size

Hortense Le Ferrand

Abstract Ceramics with heterogeneous microstructures have the potential to exhibit local variations in properties and unusual combinations of those, just like highly mineralized biomaterials do. However, to date, the microstructures achieved in technical and structural ceramics cannot rival the diversity and complexity of those found in biomaterials due to the lack of adapted processing methods. Recent research, however, demonstrated that local hardness and elastic modulus can be realized in alumina ceramics by controlling the orientation of the grains in periodically varying structures. This deliberate tuning of the grain orientation resulted from the magnetically-driven alignment of anisotropic template particles in the initial liquid suspension. During the sintering at high temperature, the template particles grew as anisotropic grains with the orientation set by the magnetic field. To expand the design freedom of such ceramics in terms of microstructure and properties, this paper aims at tuning the grain size, grain orientation, and final porosity of sintered alumina. The methodology to build multilayered ceramics using varying template particles sizes is described and examples of the microstructures obtained are provided. This work contributes to pushing forward the field of bioinspired ceramics that would hopefully give rise to structural ceramics with unusual combinations of local properties.

Keywords Microstructure · Ceramic · Grain growth · Magnetic orientation

---

H. Le Ferrand (✉)
School of Mechanical and Aerospace Engineering, School of Materials Science and Engineering, Nanyang Technological University, 50 Nanyang Avenue, Singapore 639798, Singapore
e-mail: hortense@ntu.edu.sg

B. Li et al. (eds.), *Advances in Powder and Ceramic Materials Science*, The Minerals, Metals & Materials Series,
https://doi.org/10.1007/978-3-030-36552-3_5

Introduction

Heterogeneous structures in highly mineralized materials are desired to achieve local properties, customization, and unusual combinations of functions [1]. Such unusual combinations may be, for example, damage and impact resistance with high strength and hardness [2, 3], wear-resistance and lightweightness with strength, or functional gradients. These properties contrast with most conventional ceramics that generally aim at homogeneity and are largely inspired by many biological and highly mineralized materials [4, 5]. Indeed, many hard biocomposites exhibit local variations in chemical composition, density, and reinforcement orientation. A typical example of such natural materials is the dactyl club of the Mantis Shrimp [6]. This club combines strength with impact resistance by featuring multiple gradients from the outer to the inner part: a gradient in crystallinity and mineral phase from fluorapatite to amorphous calcium phosphate, a gradient in density with an increase in organic content, and a gradient in the pitch of a helicoidal arrangement of mineral rods [7]. Reproducing such complex microstructures in structural technical ceramics remains a challenge but is desired to expand the use of ceramics. Indeed, strong, hard, and tough lightweight materials find a large range of applications in the biomedical, aerospace, and automotive industries [8, 9].

Several paths have been explored to approach nature's microstructural complexity in porous systems and reinforced composites, using methods such as field-assisted freeze casting [10] or additive manufacturing [11]. The materials fabricated by these methods, however, do not reach the high strength, hardness, and toughness of their biological counterparts. To realize such mechanical properties, the microstructural control has to be done in a highly mineralized system [12]. Recently, the use of magnetic fields to control the orientation of elongated particles in any direction in space showed promising results in designing ceramics and composites with local properties [2, 13, 14]. In one of these processes, the magnetic orientation of microparticles in liquid is combined with two traditional ceramic processes, namely, slipcasting and templated grain growth (TGG) [13]. Typically, the method consists in aligning, using a magnetic field, anisotropic template microparticles co-suspended with smaller nanoparticles of the same composition in an aqueous solution. After drying, the template particles retain their orientation and during the sintering, they grow to form large anisotropic grains along this orientation [15, 16]. Using this method, dense alumina ceramics with deliberate orientation of their grains could be fabricated. In particular, periodically varying orientations could be achieved in a variety of geometrical shapes, to reproduce the pitch gradient found in the Mantis Shrimp dactyl club [13] (Fig. 1). The local hardness and moduli were found to vary along the pitch [13], as a consequence of the dependence of crystalline orientation and number of weak interfaces along the loading direction with the mechanical properties [17, 18].

To expand the degree of design freedom in such microstructures, as well as to increase the strength and toughness, additional control over the grain size and aspect ratio is desired. Indeed, extrapolating the theory of reinforced composites to

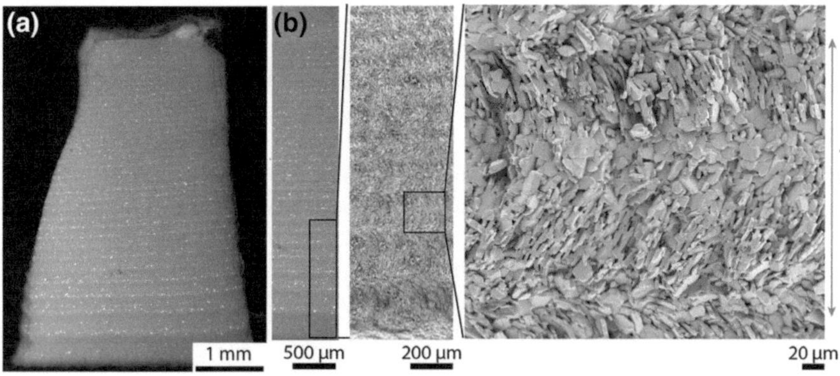

Fig. 1 Optical image of a periodically microstructured alumina ceramic (**a**) and electron microscopy images of its internal grain arrangement (**b**) prepared by magnetically-driven assembly and templated grain growth (reproduced from [13], copyright © 2019, Wiley and sons)

ceramics with aligned grains, hardness, and strength are expected to increase with the aspect ratio of the grains (Fig. 2a) [19]:

$$H_v = H_{v0} + H_{v1}\sqrt{L}, \tag{1}$$

where L is the width of the grain, H_v the hardness and H_{v0} and H_{v1} constants, and

$$\tau = \tau_0 + \tau_1\sqrt{L}, \tag{2}$$

where τ is the yield strength and τ_0 and τ_1 are constants. In addition, the fracture mode of highly textured materials changes from brittle where the grains break to pull-out mode where the grains deflect the crack at their interfaces [20]. As a result, aspect ratio along with grain dimension and orientation can be expected to influence the toughness of the material as well as the strength (Fig. 2b). Combining these

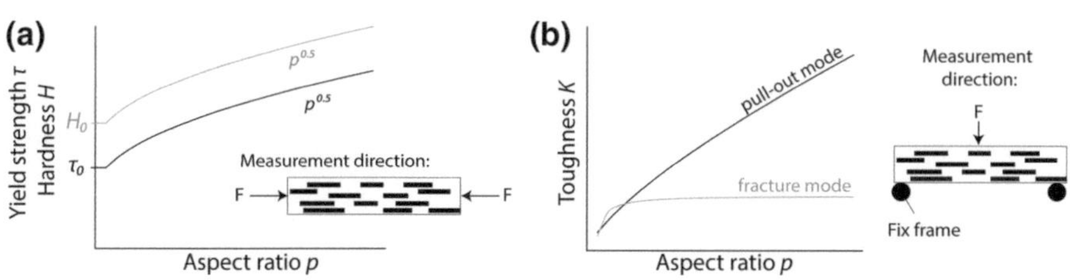

Fig. 2 Trends of the mechanical response of ceramics with anisotropic and aligned grains as a function of the aspect ratio of those grains: **a** Yield strength τ and hardness H for a load F applied along the long dimension of the grains, **b** toughness K in the case of a crack initiated perpendicularly to the alignment direction

multiple features could lead to enhanced and localized properties in structural ceramics.

To explore this design space and move one step forward in the direction of hard and tough lightweight materials, this paper describes the fabrication of multilayer alumina ceramics with controlled grain orientation and size, using the magnetically-assisted slipcasting and templated grain growth process (MASC-TGG). First, the influence of the geometrical dimensions of the initial anisotropic templates on the grain size and material porosity before and after sintering are described and compared to theoretical predictions. Then, those templates are functionalized to respond to external magnetic fields and biaxially align in arbitrary directions. Finally, an example of a multilayer ceramic with controlled grain size and orientation is presented. Such controlled microstructures could be used to design ceramic and composite materials that exhibit local properties, controlled crack path, and, therefore, enhanced strength and toughness. This work is a step forward in the design and fabrication of tough structural materials with applications in health, transport, or defense.

Materials and Methods

Slurry Preparation, Casting, and Sintering

Alumina microplatelets (Al_2O_3 pl, Kinsei Matec Co., Japan) were suspended at a ratio 1:1 in a waterborne and acidic dispersion of hydrophilic fumed alumina nanoparticles (Al_2O_3 np, AERODISP W 440, Evonik, Germany) containing 40 vol.% solids. The mixture was stirred overnight before being casted onto a gypsum substrate (Ceramix, Germany) prepared and dried earlier. After the absorption of the water from the slurry in the gypsum plaster, both sample and plaster were dried at mild temperature overnight (Binder VD 53, Fischer Scientific Pte Ltd, Singapore). The green bodies obtained were subsequently sintered in air (sintering furnace Nabertherm, Switzerland) with the following heating procedure: 5 °C/min up to 500 °C, plateau at 500 °C for 2 h, then up to 1600 °C in 6 h and plateau at 1600 °C for 2 h. The parameters for the theoretical determination of the grain growth were determined by varying the sintering temperature and applying the method described by Suvaci et al. [21].

Magnetic Response of Alumina Platelets

The alumina platelets were magnetized following a procedure described elsewhere [22]. Their magnetic response was observed under an optical microscope (Nikon Epiphot 200) in an aqueous solution of 5 wt% of polyvinyl pyrrolidone

(Sigma-Aldrich, China, MW = 360’000). The magnets used were rare-earth neodymium magnets (Eclipse, Switzerland) attached to DC motors (RS PRO, China). The images were recorded using the camera (Infinity, Lumenera) and software (Infinity) and analyzed using Image J (NIH, USA).

Characterization of the Ceramics

Green bodies and sintered ceramics were imaged using an electron microscope (FESEM 76000F, Japan). Green body samples were broken manually and coated with 5 nm of Pt. The sintered ceramics were polished and thermally etched at 1600 °C for 20 min before applying the Pt coating. The density of the samples was measured using the Archimedes principle, using toluene as a solvent (Sigma-Aldrich, India) for the green bodies and deionized water for the sintered samples.

Results and Discussion

Overview of the MASC-TGG Process

The combination of MASC with TGG was selected as the method of choice for the fabrication of ceramics with deliberate grain orientation and size (Fig. 3). Indeed, MASC-TGG is a pressure-less method that is suitable for near net-shape manufacturing [13]. The method consists in preparing a liquid suspension containing both anisotropic micron-scale particles and nanoparticles of the same chemistry, Al_2O_3 in this study. The micron-scale particles, also called platelets, are modified to respond to external magnetic fields [22]. This modification consists in adsorbing superparamagnetic iron oxide nanoparticles (Fe_3O_4 np) at their surface and has been developed elsewhere [22, 23]. When the suspension is liquid enough, the platelets can be oriented in any deliberate direction [2, 14]. Casting the suspension onto a porous mold, the liquid is slowly removed by capillary forces in a time-dependent manner [2]. As the water is removed, the viscosity rises, locking the orientation of the platelets in place (Fig. 3a). The larger particles will serve as templates to the smaller particles during the sintering at high temperature to lead to large anisotropic grain with the same geometric and crystalline orientations [15, 16]. This is the templated grain growth process that is used for the consolidation and densification of the ceramic (Fig. 3b).

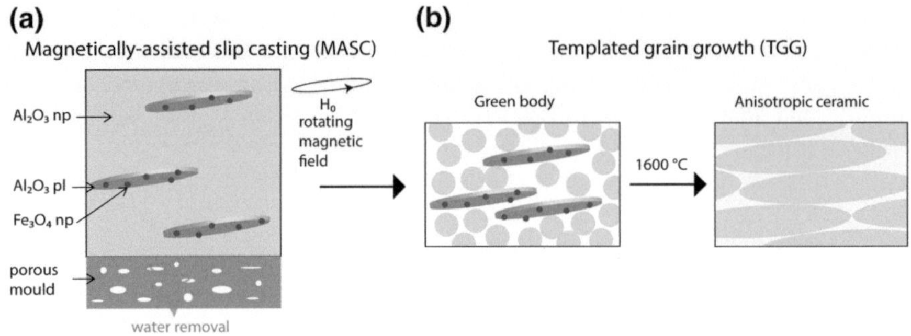

Fig. 3 Overview of the process used to fabricate the anisotropic ceramics, combining magnetically-assisted slipcasting (a, MASC) and templated grain growth (b, TGG). During MASC, magnetically responsive anisotropic alumina platelets (Al_2O_3 pl) decorated with iron oxide nanoparticles (Fe_3O_4 np) are co-suspended with alumina nanoparticles (Al_2O_3 np) in water. The platelets are oriented *via* a rotating magnetic field $\boldsymbol{H_0}$ as the liquid is removed through the pores of a mold made of plaster. The dry assembly called green body is then consolidated and densified into a ceramic by TGG at 1600 °C

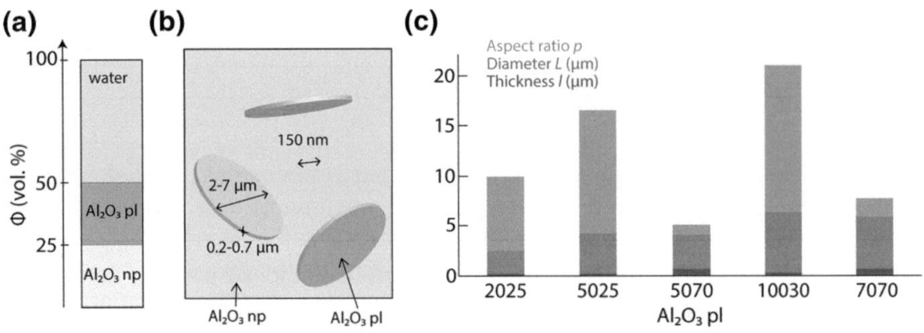

Fig. 4 **a** Composition of the slurry in terms of volume fraction, **b** schematics of the suspension and **c** geometrical dimensions of the Al_2O_3 pl used in the different slurries

Slurry Composition

The composition of the slurries was optimized for particles with the dimensions and aspect ratios used in this study (Fig. 4). The key property for the slurry preparation is to exhibit low viscosity while containing a concentration in solid high enough to maintain the particles orientations after drying [14]. To this aim, the Al_2O_3 platelets were added to a commercial suspension of alumina nanoparticles. The acidic pH around 4 allowed good dispersion. Tuning the concentration of the Al_2O_3 platelets, it was found that total solid concentrations below 50 vol.% lead to significant sedimentation of the platelets and warping of the casted samples after drying, for all the particle sizes used in this study. Furthermore, above 50 vol.% the viscosity was found to drastically increase, leading to the formation of a paste-like material.

In such high viscosity suspension, the platelets could not be aligned by a magnetic field. For the range of platelet sizes used herein the experiments, 50 vol.% total solid content was found suitable for the slurry preparation and was used in the remainder of the study. Using the same solid content for different particle sizes is convenient for the future processing of continuous gradients and to prevent distortions during drying.

Casting and Densification by TGG

The consolidation and densification of the green bodies into ceramics was achieved by templated grain growth as described earlier. The final grain dimensions and ceramic densities were found to depend on the initial platelet sizes (Fig. 5). Using the slurries containing 50 vol.% solid content, no significant deformation was observed after sintering of 3 mm-thick cylindrical samples. For all compositions, the microstructure showed elongated grains, oriented at random angles when no magnetic field is applied (Fig. 5a). The images also revealed the presence of pores suggesting that the packing of the nanoparticles around the platelets might also be affected by the aspect ratio of the platelets. The theory of the templated grain growth could be applied to the compositions to predict the final grain size (Fig. 5b), using the following equation [21]:

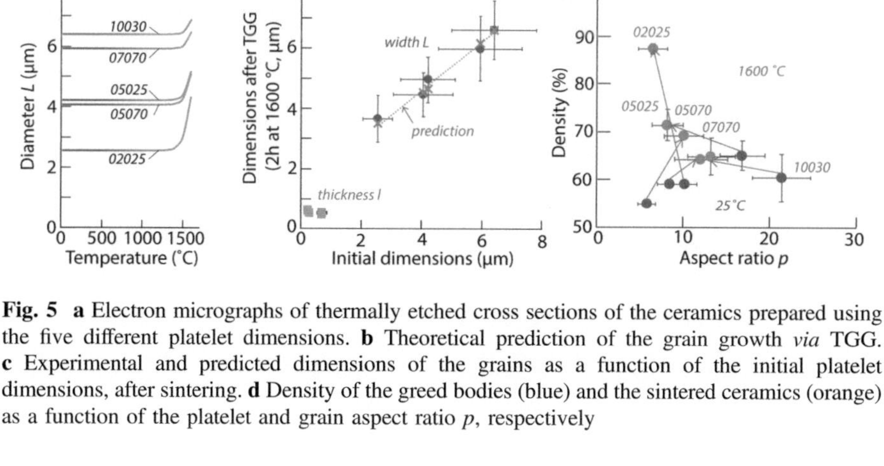

Fig. 5 **a** Electron micrographs of thermally etched cross sections of the ceramics prepared using the five different platelet dimensions. **b** Theoretical prediction of the grain growth *via* TGG. **c** Experimental and predicted dimensions of the grains as a function of the initial platelet dimensions, after sintering. **d** Density of the greed bodies (blue) and the sintered ceramics (orange) as a function of the platelet and grain aspect ratio p, respectively

$$L_s^3 - L^3 = B \cdot e^{mT}, \tag{3}$$

where L is the initial template width and L_s the width after sintering, T the sintering temperature, and B and m constants are determined experimentally: $B = 3.969 \cdot 10^{-11}$ μm^3 and $m = 0.0175$ °C^{-1}. The dimensions of the templates measured after sintering were found to follow the prediction (Fig. 5c), whereas the thicknesses of the templates were found to be of 0.5 μm approximately for all platelet dimensions. Finally, the ceramic densities also varied as a function of the initial aspect ratios of the platelets (Fig. 5d). Although there was no direct correlation between the initial aspect ratio and the final density due to the simultaneous variation of length and thickness, the final densities did show correlation with the final aspect ratios of the grains. Indeed, higher aspect ratios corresponded to lower densities. This is probably due to the lower packing density of long grains as compared with smaller and rounder grains.

Deliberate Orientation Using Magnetic Fields

The orientation of the grains in the final ceramics were controlled by magnetically orienting the platelets in the liquid slurries using rotating magnetic fields (Fig. 6) [22]. In view of using the same set-up for all compositions, the five types of platelet should be able to align for the same magnetic field strength. To this aim, their magnetic coating was modified to allow their alignment under a magnetic field of 10 mT only. First, the theory of the magnetic alignment of anisotropic particles decorated with iron oxide nanoparticles of 10 nm diameter was applied [22] (Fig. 6a). This alignment results from the competition between viscous and magnetic torques when the particle rotates along with the magnet. Thus, it was predicted that at a constant particle diameter, a higher aspect ratio—hence a higher anisotropy was favorable for magnetic alignment. In this case, the magnetic susceptibility χ required for the platelets to orient with the field decreases. Since the magnetic susceptibility is the consequence of having Fe_3O_4 nanoparticles adsorbed on the surface, the minimum concentration of Fe_3O_4 decreases with the aspect ratio. On the contrary, at constant thickness, the viscous torque dominates when the diameter of the platelet increases demanding a higher magnetic susceptibility, hence an increase in Fe_3O_4 concentration. Following those predictions, the concentration of Fe_3O_4 needed was calculated so that all platelets align at 10 mT. The obtained magnetically responsive platelets were found to exhibit the predicted alignment when suspended in a low viscosity fluid under a rotating magnetic field at ~10 mT (Fig. 6b). The microscopic images and their FFT showed a very good biaxial orientation of the platelets according to the position of the magnetic field.

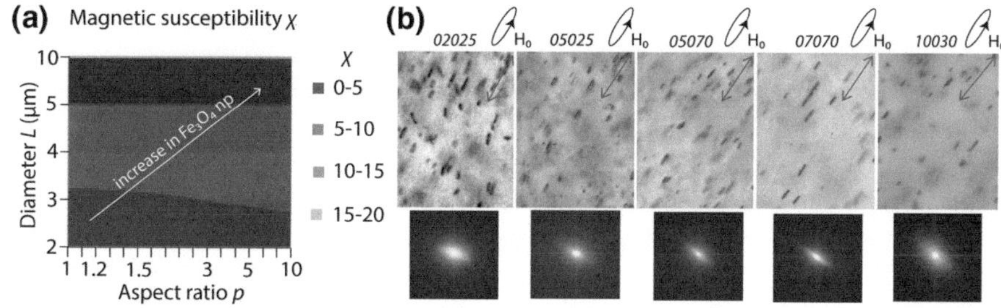

Fig. 6 **a** Theoretical determination of the magnetic susceptibility χ as a function of the diameter L of the platelet and their aspect ratio p to achieve magnetic alignment at 10 mT. The magnetic susceptibility is controlled by the concentration of Fe_3O_4 nanoparticles adsorbed at the surface of the Al_2O_3 platelets. **b** Optical images of aligned ceramic platelets suspended in low viscosity fluid and the fast Fourier transform (FFT) of the images (underneath)

Oriented and Multilayered Ceramics

MASC-TGG was applied to the slurries containing the magnetically responsive particles. The set-ups used for MASC-TGG are simple and offer a large range of variability in sample size, magnetic fields, and provide space for additional accessories such as microscope or temperature control (Fig. 7). The set-ups feature a rotating rare-earth magnet or alternatively a rotating sample, as well as the mold

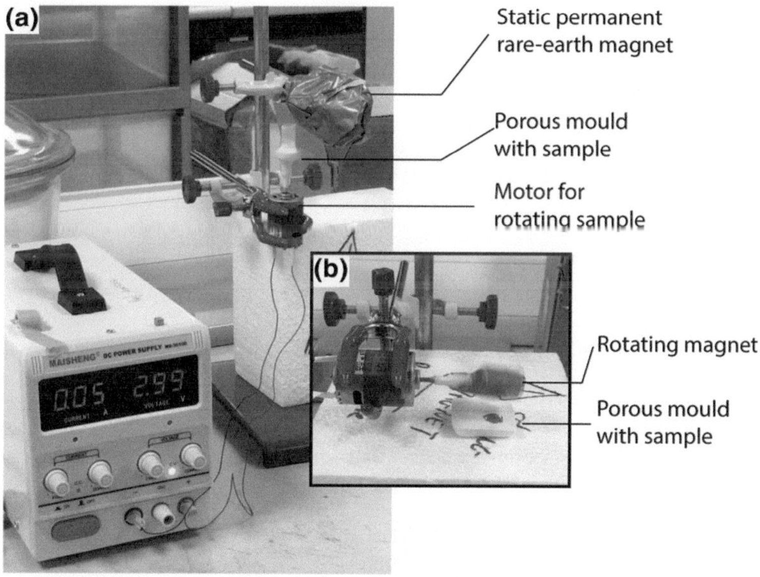

Fig. 7 Set-ups used for MASC-TGG for horizontal alignment (**a**) and for vertical alignment (**b**) of the Al_2O_3 templates

with the porous substrate. Thanks to the optimization of the slurries, the same set-ups could be used for all suspensions independently of the template size.

Using the magnetic alignment of the particles in the slurry, the green bodies as well as the sintered ceramics exhibited the intended microstructures (Fig. 8). First, it was verified that after MASC, the platelets were locked into the desired orientation set by the rotating magnetic field. This was done by choosing the least natural orientation of platelets, namely, biaxial vertical alignment (Fig. 8a). After MASC and drying of the samples, all compositions exhibited the intended vertical orientation of the platelets in the green body. The platelet alignment could be easily observed under electron microscopy after brittle fracture of the sample perpendicularly to the alignment direction. Indeed, in this case, some platelets were pulled out from the nanoparticles bed, leaving a gap that appears black under microscope. Using image analysis, the FFT of the micrographs confirmed the preferential orientation of those black features, hence a good vertical alignment (Fig. 8b). To exemplify how these slurries and the process could be used for heterogeneous bioinspired microstructured ceramic, a multilayer sample was prepared (Fig. 8c).

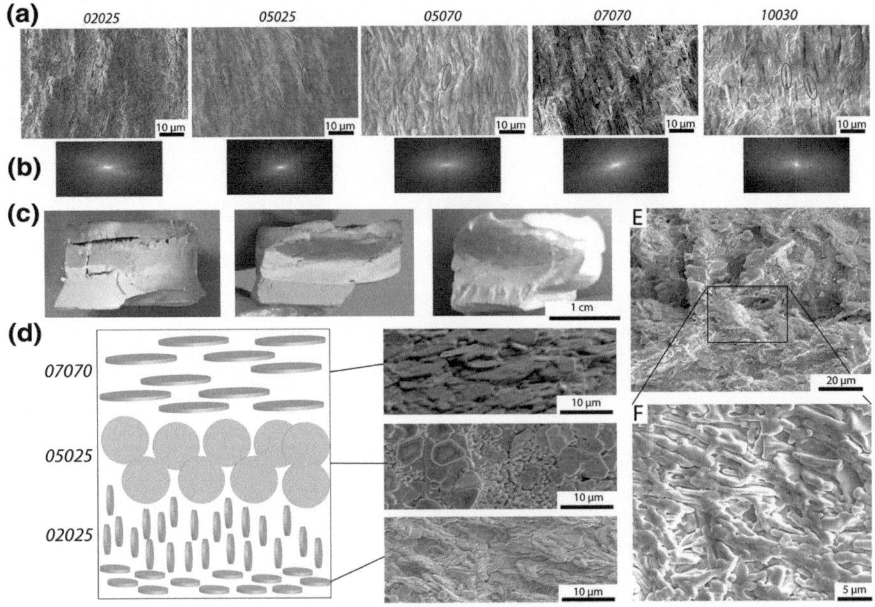

Fig. 8 **a** Electron micrographs of green bodies with vertical alignment of the Al_2O_3 platelets and the corresponding FFT **b** for the different template dimensions used. The red arrows in the images indicate the direction of the alignment of the grains, and the red ellipses circle the gaps left by platelets during the fracture revealing their orientation. **c** Pictures of a multilayer sample green bodies (left and middle) and after sintering (right). **d** Schematics of the grain orientation and particle dimensions used in the multilayer with the corresponding electron micrographs (right). Platelet contours are highlighted in yellow. View of the junction between the layers in the sintered sample before (**e**) and after polishing

In this sample, three slurries were selected, containing the platelets 02025, 05025 and 07070, and deposited from bottom to top in this order. Each of these slurries was casted sequentially after the liquid from the previous layer was evacuated by slipcasting. Due to the large thickness of each layer and the difference in magnetic coating of the platelets, the layered structure was visible optically in the green body with a change in color (Fig. 8c, left). The multilayering was more apparent after fracture of the green body in half (Fig. 8c, middle), whereas it disappeared in the sintered sample (Fig. 8c, right). The loss in color in the sintered sample can be attributed to the low concentration of iron oxide nanoparticles, whose atoms could diffuse into the alumina lattice with the high temperature sintering. Along with the change in composition, the ceramic was designed to have varying grain orientations between each layer, as schematized in Fig. 8d. It can be noted that the initial layer had two orientations. This is because slipcasting inevitably leads to the formation of a horizontally aligned layer close to the surface of the porous mold, due to the high capillary forces occurring there [14]. Corresponding electron micrographs showed the expected alignment. Finally, a closer look at the interface between the layers indicated good bonding without delamination (Fig. 8e, f). This may be attributed to the fact that after slipcasting, the sample is not entirely dried so that co-mixing by diffusion can still occur across the interface.

Conclusions

This study describes the preparation of a multilayered ceramics with simultaneous control of the local grain size, orientation, and porosity. The method employed template particles of various aspect ratios and dimensions that were modified to align in any deliberate orientation using external magnetic fields of same strength. Sequentially depositing layers containing different particle sizes and orientation, multilayered green bodies can be obtained. These green bodies were then sintered to allow for templated grain growth and densification. The final ceramic exhibited variations in local grain orientation and density, as intended by the design. Future research on this topic will involve the combination of selected suspensions with periodic microstructures of decreasing pitch, to more closely imitate the design of the Mantis Shrimp dactyl club and, hopefully, some of its outstanding properties. The ultimate motivation of the work is to obtain as strong but less brittle dense alumina ceramics.

Acknowledgements The author acknowledges financial support from Nanyang Technological University with the Start-Up grant M4082382.050).

References

1. Naleway SE, Porter MM, McKittrick J, Meyers MA (2015) Adv Mater 27:5455
2. Le Ferrand H, Bouville F, Niebel TP, Studart AR (2015) Nat Mater 14:1172
3. Le Ferrand H (2019) Compos Struct 224:111105
4. Wegst UGK, Bai H, Saiz E, Tomsia AP, Ritchie RO (2015) Nat Mater 14:23
5. Munch E, Launey ME, Alsem DH, Saiz E, Tomsia AP, Ritchie RO (2008) Science (80-.) 322:1516–80
6. Weaver JC, Milliron GW, Miserez A, Evans-Lutterodt K, Herrera S, Gallana I, Mershon WJ, Swanson B, Zavattieri P, DiMasi E, Kisailus D (2012) Science (80-.) 336:1275
7. Amini S, Masic A, Bertinetti L, Teguh JS, Herrin JS, Zhu X, Su H, Miserez A (2014) Nat Commun 5:1
8. Studart AR (2012) Adv Mater 24:5024
9. Barthelat F, Mirkhalaf M, Soc JR (2013) Interface 10
10. Nelson I, Naleway SE (2019) Integr Med Res 8:2372
11. Kokkinis D, Schaffner M, Studart AR (2015) Nat Commun 45:333
12. Le Ferrand H (2019) J Mater Res 34:169
13. Le Ferrand H, Bouville F (2019) J Am Ceram Soc 1
14. Le Ferrand H, Bouville F, Studart AR (2019) Soft Matter 15:3886
15. Seabaugh M, Messing GL (1997) J Am 88:1181
16. Pavlacka RJ, Messing GL (2010) J Eur Ceram Soc 30:2917
17. Habelitz S, Carl G, Russel C, Thiel S, Gerth U, Schnapp J-D, Jordanov A, Knake H (1997) J Non Cryst Solids 220:291
18. Suzuki TS, Sakka Y (2005) Scr Mater 52:583
19. Palmero P (2015) Nanomaterials 5:656
20. Bonderer LJ, Studart AR, Gauckler LJ (2008) Science (80-.) 319:1069
21. Suvaci E, Oh KS, Messing GL (2001) Acta Mater 49:2075
22. Erb RM, Segmehl J, Charilaou M, Löffler JF, Studart AR (2012) Soft Matter 8:7604
23. Erb RM, Libanori R, Rothfuchs N, Studart AR (2012) Science (80-.) 335:199

Compressive Properties of Micro-spherical SiO_2 Particles

Niko Hellstén, Antti J. Karttunen, Charlotta Engblom, Alexander Reznichenko and Erika Rantala

Abstract Micron-sized, spherical SiO_2 particles are important in various industrial applications, such as in heterogeneous catalyst preparation. In particular, many of industrially relevant olefin polymerization catalysts are currently prepared using micro-spherical silica as catalyst support. In large-scale catalytic polyolefin production, the quality of the final product, as well as the process efficiency is crucially dependent on overall consistency, quality, and physico-chemical properties of the catalyst. As the catalyst particle experiences various stresses during the polymer particle growth, mechanical properties of catalyst play a key role in its performance in the polymerization process. However, there is currently a lack of experimental mechanical property measurements of micron-sized, spherical SiO_2 particles relevant for the polyolefin catalyst production. In this work, compressive properties of commercial porous micro-spherical silicas were studied using a quasi-static micro-compression method. The method includes compressing single, micron-sized particles in controlled loading conditions. From the measurements, the compressive elastic–plastic properties of these particles can be determined.

Keywords Micro-compression · Micro-spherical silica · Catalysts

Introduction

Mechanical properties of solid catalyst materials are important in many industrial processes [1, 2]. In polyolefin production, the particle strength and structural integrity of heterogeneous Ziegler–Natta (ZN), Phillips, metallocene, or late transition metal catalysts directly affects the quality of polymer produced, in particular,

N. Hellstén (✉) · A. J. Karttunen
School of Chemical Engineering, Aalto University, P.O. Box 16100, Aalto, Finland
e-mail: niko.hellsten@aalto.fi

C. Engblom · A. Reznichenko · E. Rantala
Borealis Polymers Oy, PL 330, 06101 Porvoo, Finland
e-mail: alexander.reznichenko@borealisgroup.com

B. Li et al. (eds.), *Advances in Powder and Ceramic Materials Science*,
The Minerals, Metals & Materials Series,
https://doi.org/10.1007/978-3-030-36552-3_6

affecting fragmentation and generation of undesired polymer fines in the polymerization process [3, 4]. The commonly used heterogeneous ZN catalysts are porous spheres or quasi-spheres of 10–100 μm in diameter, and comprise a transition metal component, organometallic co-catalyst and are produced using $MgCl_2$ or SiO_2 support. Consequently, experimentally measured mechanical strength data of the respective catalyst support materials, such as micro-spherical, porous SiO_2 particles can provide important insight for selecting adequate catalyst support for process optimization [1].

Strength of materials depends on their density, porosity, and microstructure (crystallinity, flaws, flaw distribution) and dictates their responses to external mechanical forces [5]. These responses include deformation and fracture characteristics under uniaxial or multiaxial stresses. Based on their strain behavior and fracture type, materials are either brittle or ductile. Ceramic materials, such as Al_2O_3 and SiO_2, are generally brittle and exhibit mainly elastic deformation prior to fracture [6].

Spherical shape of catalyst particles limits the study of their mechanical properties. Thus, compression tests ('Crushing' tests) and impact tests, rather than tensile tests, have been used to study strengths of spherical catalyst carrier materials, both in macro- (>1 mm) and micron-size (<1 mm) [2, 7]. These previous studies confirm the brittle nature of Al_2O_3 and SiO_2 and reveal a scatter in the measured breaking loads and strengths. This scatter is due to the initiation of failure in a randomly distributed flaw, possibly with less force than the material strength. Consequently, measurement of the intrinsic material strength is difficult, and statistical methods such as Weibull distribution analysis are applied to better estimate material failure.

While experimental measurements of the strength of silica exist, experimental data is still lacking for porous micro-spherical silica. In this work, the compressive properties of micro-spherical, porous 16.5–49.4 μm diameter silica particles were studied experimentally using the micro-compression method. This method is similar to the American Society for Testing and Materials (ASTM) compression test for samples with dimensions greater than 1 mm [8]. From the measurements, the breaking strengths of single particles were calculated.

Materials and Methods

Table 1 shows the properties of the tested porous, micro-spherical materials. Due to assumed sample brittleness, a minimum of 40 measurements of each sample was conducted to increase statistical reliability of the results. True densities of the samples were determined via He gas pycnometry method, using [9] AccuPyc 1330 instrument from Micromeritics.

All of the micro-compression measurements were conducted using a Shimadzu MCT-511 micro-compression tester. Figure 1 depicts a schematic of the setup. It consists of a computer-controlled stage, upon which the lower compression plate is

Table 1 Properties of the tested silicas

Sample	# of individual particle measurements	Diameter range (μm)	Average size (μm)
A	40	23.4–49.4	34.9
B	42	16.5–45.9	28.3

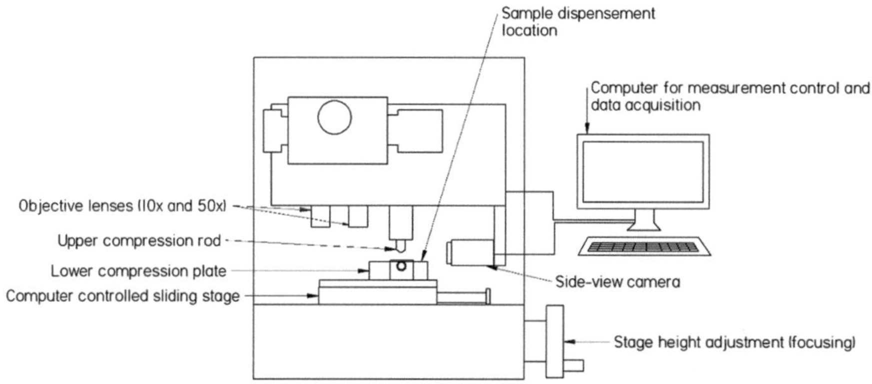

Fig. 1 Schematic of the micro-compression setup

fastened, optical microscope for sample positioning, and force-controlled upper compression head. The displacement and force measurement sensors are located at the upper part of the tester in close contact with the upper compression head.

Table 2 shows the conditions of the micro-compression measurements.

As the particle sizes ranged from 16.5 to 64.3 μm, a 50 μm diameter flat tip diamond indenter was used. In the beginning of a measurement set, the surface of the lower compression plate was cleansed with alcohol, dried, and fastened to place. Next, the sample particles were dispensed onto it using microspatulas. Typically, in one dispensing several particles ended on the lower compression plate (Fig. 2a). When the sample was in place, the measurement parameters were set using the control and analysis software provided with the device. For the measurements reported here, the compression mode was used. In this mode, the sample is compressed with a fixed loading rate until the set test force value is reached, and then held for a set time after which the test ends by the compression head returning to its original position. Other test modes for this equipment are cyclic and load–unload. For these measurements, the sample breakage was chosen as the end condition. This means that if the particle breaks before reaching maximum force, the measurement ends.

Test force of 20 mN and a slow loading rate of 0.4462 mN/sec were chosen based on initial test measurements at higher forces and faster rates, which resulted in particles flying away from the lower compression plate. Suitably isolated single particles for the measurement were located using the optical microscope and

Table 2 Conditions of the micro-compression measurements

Indenter type	Flat indenter with a 50 μm diameter diamond tip
Objective lens for size measurement	50×
Measure mode	Single
Test mode	Compression
Test form	Particle
Test end condition	Breaking point
Test force (mN)	20
Loading rate (0.4462 mN/s)	10.0
Holding time at load (s)	5
Compression ratio	10%

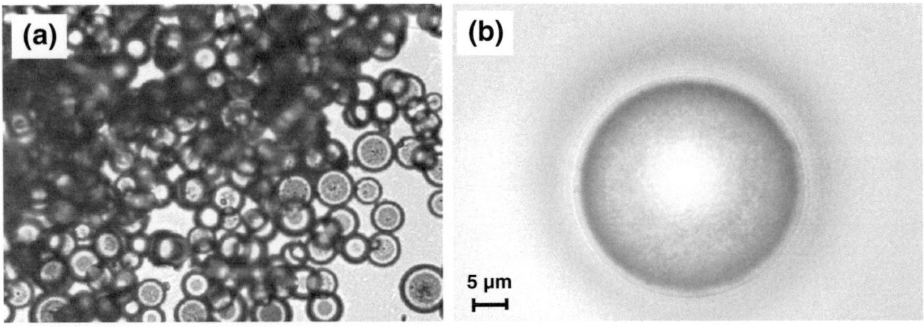

Fig. 2 **a** Multiple particles too close to each other and **b** a particle focused for size measurement and compression

computer-controlled motorized stage of the micro-compression tester. A specimen suitable for testing must be located far enough from any other particles so that only the selected particle is compressed during measurement. In case where all particles were located too close to each other, they were dispersed using a small blow of compressed air. When a suitable particle was found, its diameter was measured via digital image analysis, taking the image using 50× objective lens in the optical microscope focused onto the equatorial level of the sample. After sample size measurement, the microscope was focused on top of the sample to prevent the compression head from hitting the samples or stage, the stage was moved from the observation side to the indentation side and compression initiated. Monitoring the sample position during compression measurement is possible from one direction using the side-view camera included in the device.

In the compression measurement, the compression head is lowered until it detects the sample surface. Then the compression force increases according to the set rate. The sample surface is detected from the increase in resistance confronted by the compression head. The breaking force of each sample, displacement of the compression head from the surface, breaking and reference strengths are reported.

Displacement and force are directly measured while breaking and reference strengths are calculated according to Eq. 1. The reference strength is the calculated pressure required to deform the sample to a set value of its original size. In the measurements reported here, this value was set to 10%.

Results and Discussion

In this study, two types of commercially available porous micro-spherical silicas were investigated by uniaxial quasi-static micro-compression method. The crushing strengths of 40 particles of silica type A and 42 particles of silica type B were investigated by uniaxial quasi-static micro-compression method. The number of particles tested was selected in order to achieve statistically significant results (Table 1). Each tested particle broke during the compression sequence. Table 3 shows the breaking load, breaking stress, and stress at 10% deformation of the measurements. True densities of the sample A and B were 2.25 and 2.26 g/cm^3, respectively, as measured by gas pycnometry.

During testing, the force and displacement values are directly measured. Nevertheless, as the breaking force and displacement (deformation) are size-dependent, the more conventional stress–strain curves have been calculated using Eqs. 1 and 2, respectively. Figure 3 shows typical force–displacement and stress–strain curves of the two samples. Equation 1 was introduced by Hiramatsu and Oka [10] to calculate the ultimate strength of spherical samples in compression. Equation 2 is the conventional calculation of linear strain. In this case, only the

Table 3 Statistical values of the compressed SiO_2 samples

Sample	A	B
Load (mN)		
Mean	5.04	1.80
Median	4.84	1.63
Standard deviation	1.49	0.74
Range of values	1.85–8.71	0.33–4.08
Breaking stress (MPa)		
Mean	3.30	1.81
Median	3.33	1.88
Standard deviation	0.85	0.53
Range of values	1.79–6.45	0.27–2.83
Stress at 10% (MPa)		
Mean	1.72	1.53
Median	1.70	1.64
Standard deviation	0.70	0.61
Range of values	0.31–4.47	0.27–2.58

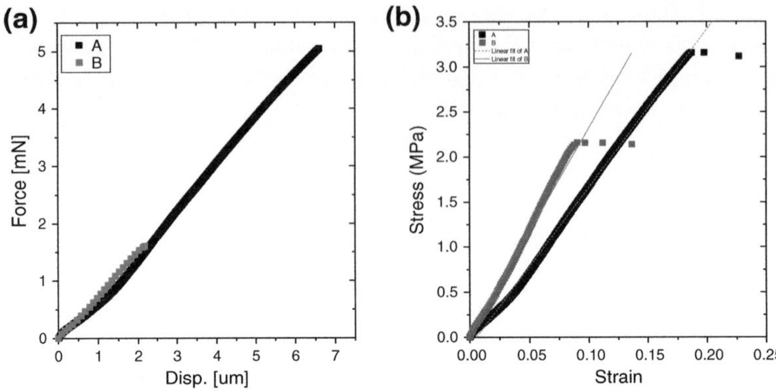

Fig. 3 Typical force–displacement (**a**) and stress–strain (**b**) curves for the studied samples

linear strain along the compression axis, e.g., the flattening of the sample, is considered (Fig. 5).

$$\sigma = 2.8 \times \frac{P}{\pi d^2} \tag{1}$$

$$\varepsilon = \frac{Measured\ deformation}{Original\ diameter\ of\ sphere} \tag{2}$$

In Eqs. 1 and 2, σ is the breaking strength (MPa), P the measured breaking force (N), d the diameter of the sample (mm), and ε the axial strain. Typically, a negative sign is used for compression measurements as the length of a sample actually decreases. In this work, it was decided to use positive values for strain for easier interpretation of the curves.

Figure 3 shows that both samples exhibit typical brittle material behavior. The curves are almost straight lines until the point of breakage, indicating mainly an elastic deformation. For both samples, there is a small change in the angle of the curve at 0.03 strain for the silica A and 0.02 strain for silica B. After these points, the curve gets steeper. This angle can indicate a yield point of the sample or small internal breakage and elastic deformation before the sample enters the elastic–plastic region [2, 11]. The variance in the relative axial deformations in these experiments was large, ranging from 7.13% to 67.70%.

The mean values indicate that the silica type A is stronger towards crushing in comparison to type B. The median values of the measurements for both samples are close to the mean values suggesting that the mean is a good measure of the measured samples. Nevertheless, the range of measured breaking force and calculated strength values are larger than the mean values indicating considerable scatter. Figure 4 represents the arrangement in compression measurement with contact points between the upper flat indenter and sample marked with A, and contact point between the sample and lower compression plate marked with B. The small circles

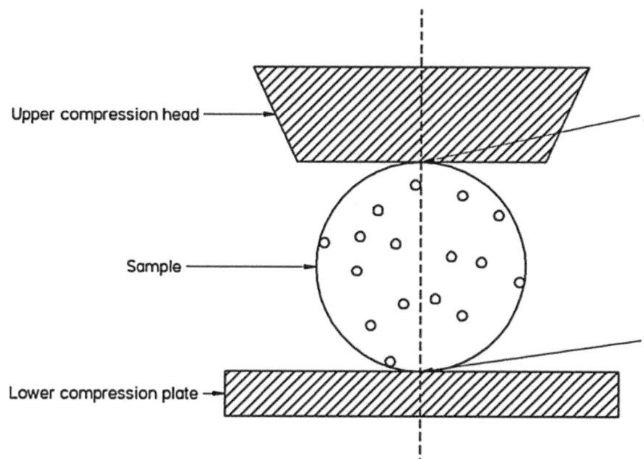

Fig. 4 Schematic of particle compression between two flat plates. A and B mark the initial contact points while the dashed line denotes the compression axis. The small circles represent randomly distributed flaws

represent the randomly distributed flaws. As particle failure can initiate in any of the flaws and with a lesser stress than the intrinsic material strength would require, a scatter in compressive strength measurements for brittle materials is typical. While the classical theories state that the strength of a material is constant throughout its volume, experimental measurement of this is extremely difficult. Nevertheless, variable results can be due to anisotropy of a material. In addition, sample volume affects strength properties and the size of the tested sample should always be reported. Due to the inability of the classical theories of strength to explain the scatter, the statistical theory of Weibull [12] is often applied in predicting probability of failure under a specified stress.

The two-parameter Weibull distribution equation is

$$F(\sigma) = 1 - \exp(-\beta_0 \sigma^m) \tag{3}$$

where $F(\sigma)$ is the probability of failure, σ the maximum stress within the sample, β_0 a size parameter, and m the Weibull modulus. The Weibull parameters can be obtained by first reorganizing Eq. 5 and taking natural logarithms from both sides to give

$$lnln(\frac{1}{1 - F(\sigma_i)}) = mln\sigma + ln\beta_0 \tag{4}$$

To obtain the probability of failure for each particle, the strength failure data of each tested particle is organized in ascending order and ranked accordingly. An estimated cumulative probability of failure is then given by

$$F(\sigma_i) = \frac{i - 0.5}{N} \tag{5}$$

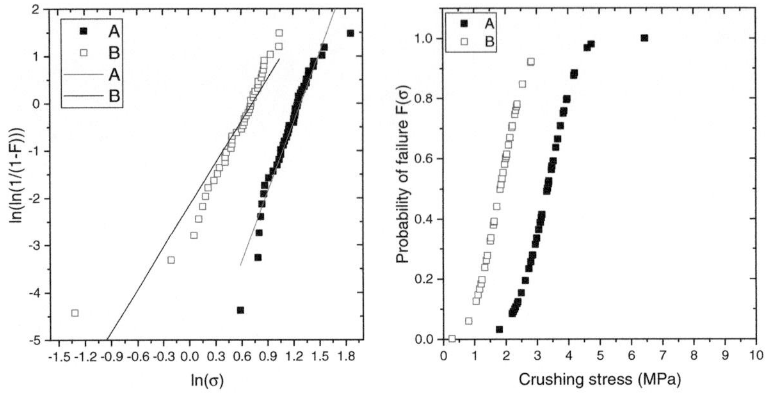

Fig. 5 Weibull probability (**a**) and survival (**b**) plot

The probability of failure and failure strength of each tested particle are used to obtain the Weibull parameters β_0 and m by linear regression. This approach was adopted from Subero-Couroyer et al. [13]. Figure 5 shows the Weibull probability and the survival plots for the tested materials. The Weibull probability plot shows how well a set of data follows the Weibull distribution. In contrast, the Weibull survival plot describesthe probability of particle failure under a specific stress.

Figure 5 shows that silica A has a higher probability to survive at greater stresses than B. In contrast, silica B results (R^2 = 0.93) fit better to the Weibull distribution in comparison to the results from type A (R^2 = 0.90). The Weibull modulus and σ_0 for A are higher than for B suggesting a stronger, more homogeneous material (Tables 4 and 5).

Comparing the results of this work with previous compression measurements of micro-spherical SiO_2 shows that the breaking strengths obtained in this work are two orders of magnitude smaller than those measured previously for fused amorphous SiO_2. Porosity may explain these differences as in comparison, Ryshkewitch [14] found that increasing porosity in Al_2O_3 can decrease its strength by two orders of magnitude from ~1000 MPa to 10–20 MPa.

Table 4 Parameters obtained from Weibull analysis

Weibull analysis	A	B
Weibull modulus, m	4.9	3.0
β_0	1.87×10^{-3}	0.12
Scale parameter, σ_0 (MPa)	3.6	2.1
R^2	0.903	0.928

Table 5 Previously measured experimental data for micro-spherical SiO_2

Sample	Diameter (μm)	Breaking strength (MPa)	Yield strength (GPa)	Young's modulus (GPa)	References
Fused amorphous	20–60	900			[2]
Sicastar plain (amorphous)	0.5		6.1	44.7	[15]
Amorphous	0.5–0.8			30.8	[16]
TYPE B	16.5–45.9	1.81			This work
TYPE A	23.4–49.4	3.30			This work

Conclusions

In this work, micro-compression method was successfully used to measure the breaking strengths of two different types of microporous SiO_2 particles. For statistical reliability, 40 and 42 individual particles of the materials were measured, respectively. Both materials exhibited brittle fracture behavior and large scatter in the results, as expected. Statistical Weibull method was applied to the results to estimate breaking probabilities of the materials. Silica A was found to be mechanically stronger than silica grade B. Based on the Weibull analysis, 92% of silica B fail at 2.8 MPa compressive stress while 100% of A fail at 6.5 MPa compressive stress. As the true densities of the samples measured by gas pycnometry are similar, these differences are likely to originate from the morphological differences such as porosity, pore connectivity, and pore size distribution. In addition, differences between previous studies of micro-spherical fused amorphous SiO_2 and this work suggest a strong effect on compressive strength due to porosity and further studies are required to confirm this.

References

1. Wu D, Zhou J, Li Y (2007) Mechanical strength of solid catalysts: recent developments and future prospects. AIChE J 53:2618–2629
2. Pejchal V, Žagar G, Charvet R, Dénéréaz C, Mortensen A (2017) Compression testing spherical particles for strength: theory of the meridian crack test and implementation for microscopic fused quartz. J Mech Phys Solids 99:70–92
3. Soares JBP, McKenna TF (2012) Polyolefin reaction engineering. Wiley-VCH; Wiley [distributor], Weinheim: Chichester
4. Hoff R, Mathers RT (eds) (2010) Handbook of transition metal polymerization catalysts. Wiley, Incorporated, Hoboken
5. Hosford WF (2009) Mechanical behavior of materials. Cambridge University Press, New York
6. Pelleg J (2014) Mechanical properties of ceramics. Springer, Cham

7. Couroyer C, Ning Z, Ghadiri M, Brunard N, Kolenda F, Bortzmeyer D, Laval P (1999) Breakage of macroporous alumina beads under compressive loading: Simulation and experimental validation. Powder Technol 105:57–65
8. Hemker KJ, Sharpe WN (2007) Microscale characterization of mechanical properties. Annu Rev Mater Res 37:93–126
9. Tamari AAS (2004) Optimum design of the variable-volume gas pycnometer for determining the volume of solid particles. Meas Sci Technol 15:1146
10. Hiramatsu Y, Oka Y (1966) Determination of the tensile strength of rock by a compression test of an irregular test piece. Int J Rock Mech Min Sci Geomech Abstr 3:89–90
11. Khanal M, Schubert W, Tomas J (2008) Compression and impact loading experiments of high strength spherical composites. Int J Miner Process 86:104–113
12. Weibull W (1939) A statistical theory of the strength of materials, Ingenioersvetenskapsakad. Handl. No. 151:45
13. Subero-Couroyer C, Ghadiri M, Brunard N, Kolenda F (2003) Weibull analysis of quasi-static crushing strength of catalyst particles. Chem Eng Res Des 81:953–962
14. Ryshkewitch E (1953) Compression strength of porous sintered alumina and zirconia. J Am Ceram Soc 36:65–68
15. Romeis S, Paul J, Ziener M, Peukert W (2012) A novel apparatus for in situ compression of submicron structures and particles in a high resolution SEM. Rev Sci Instrum 83:095105
16. Paul J, Romeis S, Tomas J, Peukert W (2014) A review of models for single particle compression and their application to silica microspheres. Adv Powder Technol 25:136–153

Part IV
Ceramic-Based Composite Materials

Microwave Absorption Properties of Polymer-Derived SiCN(Fe)/Si_3N_4 Ceramics

Xiao Lin, Hongyu Gong, Yujun Zhang, Jianqiang Bi, Yurun Feng and Shan Wang

Abstract Iron-doped SiCN/Si_3N_4 ceramics were synthesized by dipping-pyrolysis polysilazane with iron(III) acetylacetonate and use porous silicon nitride as matrix. The polysilazane pyrolyzed a certain amount of free carbon and silicon nitride at the high temperature. The content of iron increased with the increase in the amount of iron(III) acetylacetonate addition. When the sample was added with 5wt% iron(III) acetylacetone, the ε' reached the lowest value of 1 at 15.4 GHz, the highest peak value of the tan ε was 3.09 at 15.2 GHz, and the R of the sample was lower than −20 dB at 15.5 GHz, which implied that it has great wave absorption performance in the high-frequency region and could absorb more than 99% of the incident electromagnetic waves. The uniform distribution of C, β-SiC, and α-Fe in the samples is helpful to improve the wave absorption properties of the materials.

Keywords Porous Si_3N_4 · Iron(III) Acetylacetonate · Polysilazane Dipping-pyrolysis · Wave-absorbing properties

---

X. Lin · H. Gong (✉) · Y. Zhang · J. Bi · Y. Feng · S. Wang
Key Laboratory for Liquid-Solid Structural Evolution & Processing of Materials of Ministry of Education, Shandong University, Jinan 250061, China
e-mail: hygong@sdu.edu.cn

X. Lin
e-mail: linxiao050994@sina.com

Y. Zhang
e-mail: yujunzhangcn@sdu.edu.cn

J. Bi
e-mail: bjq1969017@163.com

Y. Feng
e-mail: 765132878@qq.com

S. Wang
e-mail: 1958528628@qq.com

X. Lin · H. Gong · Y. Zhang · Y. Feng · S. Wang
Key Laboratory of Special Functional Aggregated Materials, Ministry of Education, Shandong University, Jinan 250061, China

B. Li et al. (eds.), *Advances in Powder and Ceramic Materials Science*,
The Minerals, Metals & Materials Series,
https://doi.org/10.1007/978-3-030-36552-3_7

Introduction

Absorbing material is a kind of electromagnetic functional material which can effectively absorb incident electromagnetic wave, consume electromagnetic energy into heat energy or other forms of energy, or eliminate electromagnetic wave interference, so that the echo intensity of the target is significantly weakened [1, 2]. With the development and progress of science and technology and the demand of the contemporary national industry, the corresponding invisible technology and electromagnetic shielding technology of absorbing materials are becoming more and more mature, and their importance is becoming more and more prominent. In the military field, by coating the wave-absorbing material on the military equipment such as military vehicles, the reflected signal is attenuated and the detection wave is absorbed, so that the equipment is invisible and the purpose of anti-radar reconnaissance is achieved. At the same time, the electromagnetic shielding property of the wave-absorbing material can also reduce the external interference of the radar and the communication equipment, so as to improve the sensitivity and the call quality. In daily life, the electromagnetic shielding characteristics of absorbing materials can also avoid the harm caused by the electromagnetic radiation of daily electrical equipment [3–6].

Three physical processes, reflection, absorption, and transmission, occur when the electromagnetic wave shines on the surface of the material. Two basic principles to be satisfied in ideal absorbing materials [7]: a. Impedance matching principle: the impedance of the surface material should be matched with the impedance of the air in order to reduce the direct reflection of the electromagnetic wave on the surface of the material and make the electromagnetic wave enter the material to the maximum extent [8]. b. Attenuation principle: the material should have the high attenuation ability to the electromagnetic wave, once the electromagnetic wave enters the material inside, can be attenuated by the material quickly and almost all, in order to reduce the energy of the transmitted electromagnetic wave [9]. However, no single material can meet these two ideal conditions at the same time.

The porous silicon nitride has low density, good mechanical properties, small dielectric constant, and good heat resistance, so it can be used as the matrix of permeable material or high-temperature absorbing material to meet the principle of impedance matching. Polymer precursor ceramics are a kind of structural/functional integrated materials, which can decompose free carbon, SiC, Si_3N_4, and other absorbing particles through polymer precursors which can be uniformly dispersed in the matrix to realize the transformation of organic polymer precursors to inorganic ceramics, and prepare wave-absorbing multiphase ceramics that meet the attenuation principle. If the magnetic elements such as Fe, Co, or Ni are introduced by controlling the preparation process, the wave absorption loss can be further increased and the wave absorption properties of composite ceramics can be improved.

In recent years, the composites of magnetic loss and dielectric absorbing materials have become the focus of research because of their good absorbing properties, but there is little study on the electromagnetic absorbing properties of PDCs ceramic and porous ceramic matrix composites. Quan Li et al. [10] use polysilazane as precursor and porous Si_3N_4 as preform, $SiCN/Si_3N_4$ composite ceramics were successfully prepared by PIP method. The phase composition of SiCN ceramics and the electrical and dielectric properties of $SiCN/Si_3N_4$ composite ceramics were studied after annealing at the temperature of 900 to 1400 °C. The dielectric loss of the sample is improved due to the appearance of the N-doped SiC nanocrystals after high-temperature annealing. Liu et al. [11] infiltrated SiCN into porous Si_3N_4 ceramics with different precursor gas flux ratios by low-pressure chemical vapor deposition/infiltration, and $SiCN/Si_3N_4$ ceramics were prepared. Free carbon is uniformly dispersed in amorphous and low-conductivity SiCN, which makes the dielectric constant of the material increase properly, and then the dielectric properties of the material are improved.

In this study, polymer-derived $SiCN(Fe)/Si_3N_4$ ceramics were prepared by dipping-pyrolysis polysilazane with 1–7 wt% iron(III) acetylacetonate. Porous Si_3N_4 ceramic with porosity of 45% was used as matrix. The effect of iron(III) acetylacetonate addition on the wave absorption properties of the composite ceramics was investigated.

Experimental Procedure

Iron(III) acetylacetonate (purity > 98%, Shanghai Macklin Biochemical Technology Co., Ltd PR China) and polysilazane (liquid, Guangzhou Honghai Chemical Technology Co., China) were used as received. Mix the two of them until the powder dissolves completely and then used as the precursor. Porous Si_3N_4 Ceramic with porosity of 45% was used as matrix. The preparation process of porous Si_3N_4 ceramic was described elaborately in my other paper [12]. Immerse the porous Si_3N_4 matrices in the mixed precursor solution and vacuum for 20 min at the same time. Then put the preforms into a pipe furnace and pyrolyze at 1000 °C for 2 h under the atmosphere of nitrogen. Repeat the immersion-pyrolysis cycle until the qualities of the samples do not increase anymore.

X-Ray diffraction (XRD, EVO-18, CARL ZEISS SMT Ltd) was used to analyze the composition of the samples. The microstructures were observed by scanning electron microscope (SEM, JSM-6380LA). The basic electromagnetic parameters were measured through coaxial transmission reflection method using vector network analyzer (VNA, 5244A). The magnetic performance was tested by vibration sample magnetometer (VSM, JDAW-2000D).

Results and Discussion

X-ray diffraction patterns of SiCN(Fe)/Si_3N_4 ceramics with 1–7 wt% iron(III) acetylacetonate addition are presented in Fig. 1. The composition of the porous silicon nitride matrix is showed for comparison. The matrix consisted of Si_3N_4 and $Y_2Si_3O_3N_4$. The Y–Si–O–N compound was generated during the porous silicon nitride sintering process between sintering aids. And $Y_2Si_3O_3N_4$ still played a role in promoting the reaction during the next pyrolysis procedure. As can be seen from the diagram, the contents of C, β-SiC, and α-Fe increased gradually in the composite ceramics with more than 3 wt% iron(III) acetylacetone. The free carbon and β-SiC were formed by pyrolysis of polysilazane. The α-Fe is reduced by iron acetylacetone at high temperature.

When the electromagnetic wave is incident into the material, there will be three reactions, reflection, absorption, and transmission. The porous silicon nitride matrix can reduce the direct reflection of the electromagnetic wave on the surface of the material, and enable the electromagnetic wave to enter the inside of the material to the maximum extent. The α-Fe will cause magnetic loss, free carbon and β-SiC will produce dielectric loss in the material, in order to reduce the energy of the transmitted electromagnetic wave. Therefore, the increase of C, β-SiC, and α-Fe content is beneficial to the improvement of wave-absorbing property.

Figure 2 shows the SEM images of SiCN(Fe)/Si_3N_4 specimens with different additions of iron(III) acetylacetonate on fracture surfaces. When we added 1 wt% iron(III) acetylacetonate, we can see a number of holes with different sizes in Fig. 2a. The pores in Fig. 2b were reduced and particles of different sizes appeared inside. This is because more polysilizane pyrolysis and reducing iron are filled in the pores of the matrix. When 5 wt% of iron(III) acetylacetonate is added, the inside of the sample is more smooth, the arrangement of the particles is tight, the

Fig. 1 XRD patterns of SiCN(Fe)/Si_3N_4 ceramics with different iron(III) acetylacetonate addition

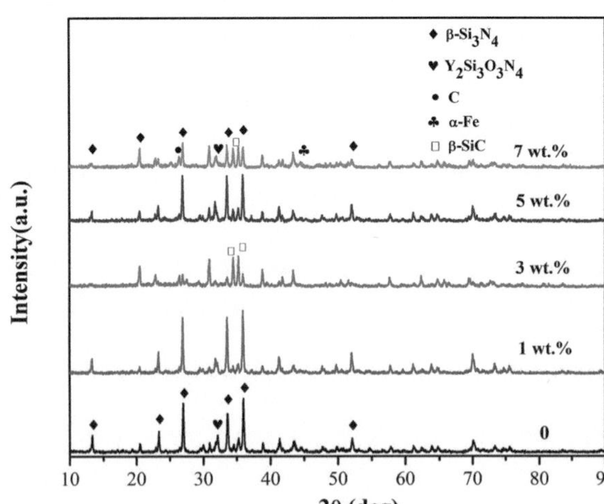

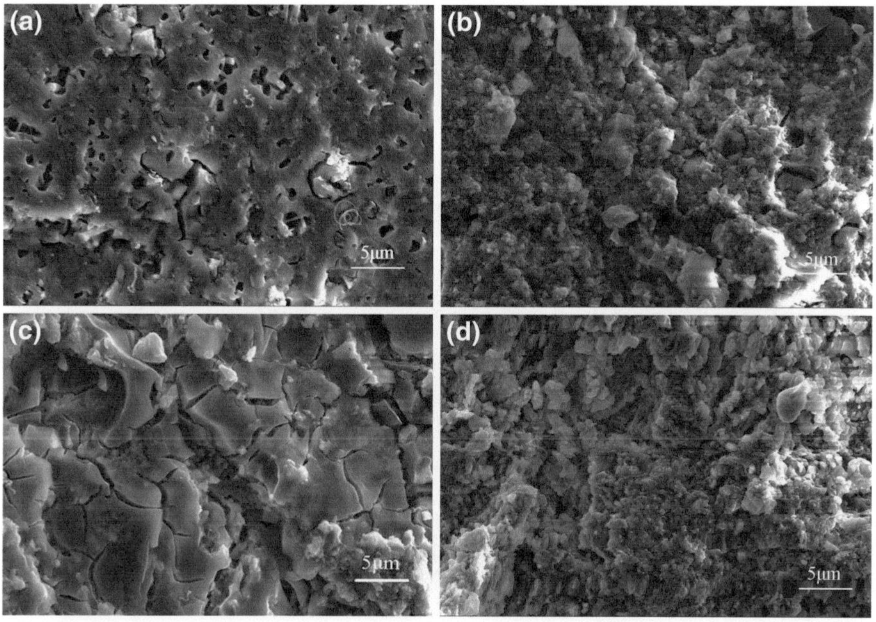

Fig. 2 SEM images of $SiCN(Fe)/Si_3N_4$ specimens with different addition of iron(III) acetylacetonate on fracture surfaces. **a** 1 wt%; **b** 3 wt%; **c** 5 wt%; **d** 7 wt%

electromagnetic wave absorption and the mechanical property of the composites can be improved as well. When the addition of iron(III) acetylacetonate continued to be increased, the cross section of the sample became very rough, which indicated that too many newly formed substances have emerged and that may be unfavorable to the propagation and loss of electromagnetic wave in the sample.

EDS analysis of one of the areas of $SiCN(Fe)/Si_3N_4$ specimens with 5 wt% iron (III) acetylacetonate is shown in Fig. 3. The specific distribution of the four main elements in the sample is shown in Fig. 3a. And we can see their relative contents in Fig. 3b. Silicon and nitrogen are found in large quantities in the sample. Although the relative contents of carbon and iron are relatively small, they are uniformly dispersed in the sample. From Table 1, we know that the amount of silicon contained in this region is 48.66 wt%, which accounts for 31.22% of atomic amount. The atomic amount of nitrogen in the area is 48.80%. The ratio of the atomic amounts is close to 3/4 in Si_3N_4. According to XRD analyzed above, we speculate that some nitrogen elements that do not form silicon nitride are used to form $Y_2Si_3O_3N_4$, and carbon element exists in the form of free carbon. In addition, we can infer that the iron content in the sample is very small.

The electromagnetic properties and wave-absorbing performance of the composite ceramics are shown in Fig. 4. In the applied magnetic field, μ' is the quantization unit of the magnetization, μ'' is the quantization unit of the magnetic loss, and the magnetic loss tangent $\tan\mu$ is a measure of the energy conversion size

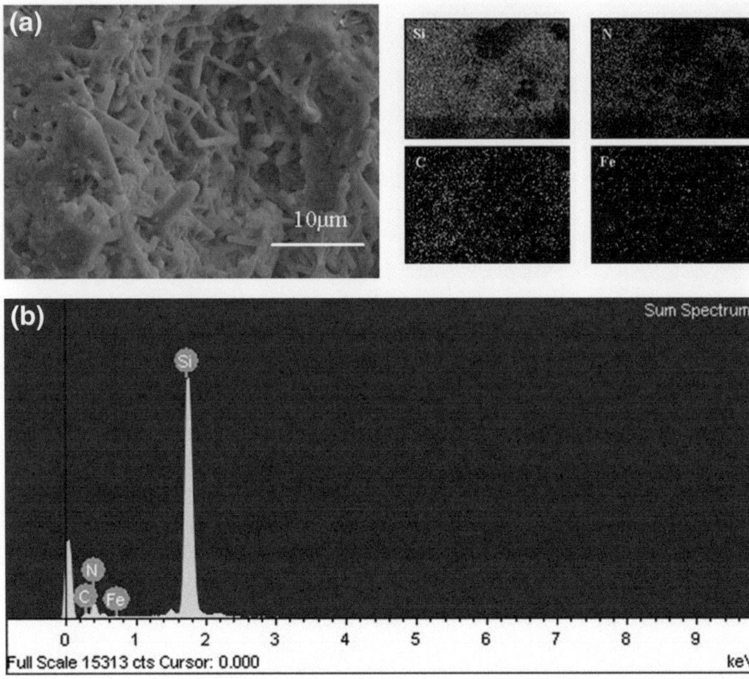

Fig. 3 EDS analysis of SiCN(Fe)/Si_3N_4 specimens with 5 wt% iron(III) acetylacetonate (**a** and **b**)

Table 1 Element analysis of SiCN(Fe)/Si_3N_4 specimens with 5 wt% iron(III) acetylacetonate

Element	Weight/%	Atomic/%
Si	48.66	31.22
N	37.93	48.80
C	13.30	19.95
Fe	0.10	0.03
Totals	100.00	100.00

of the electromagnetic wave. Similar to the results of dielectric parameters, the magnetic parameters of the samples are stable in 2–12 GHz, the real part of permeability is in the range of 0.5–1.5, and tan μ is kept between 0 $\sim$ −1. In the range of 12–18 GHz, the sample parameters of 3 wt% and 7 wt% acetylacetone iron remain stable, and the resonance peaks of 1 wt% and 5wt% samples appear at the corresponding position with dielectric parameters, which is related to the induced magnetic field in the sample [13]. It can be seen from the magnetic hysteresis loops diagram that the saturation magnetization strength of the samples containing 7 wt% acetoacetone iron is significantly higher than that of the other three samples, which is due to the highest iron content. All the magnetic hysteresis loops of the samples remain intact, but the residual magnetization and the coercivity are very small,

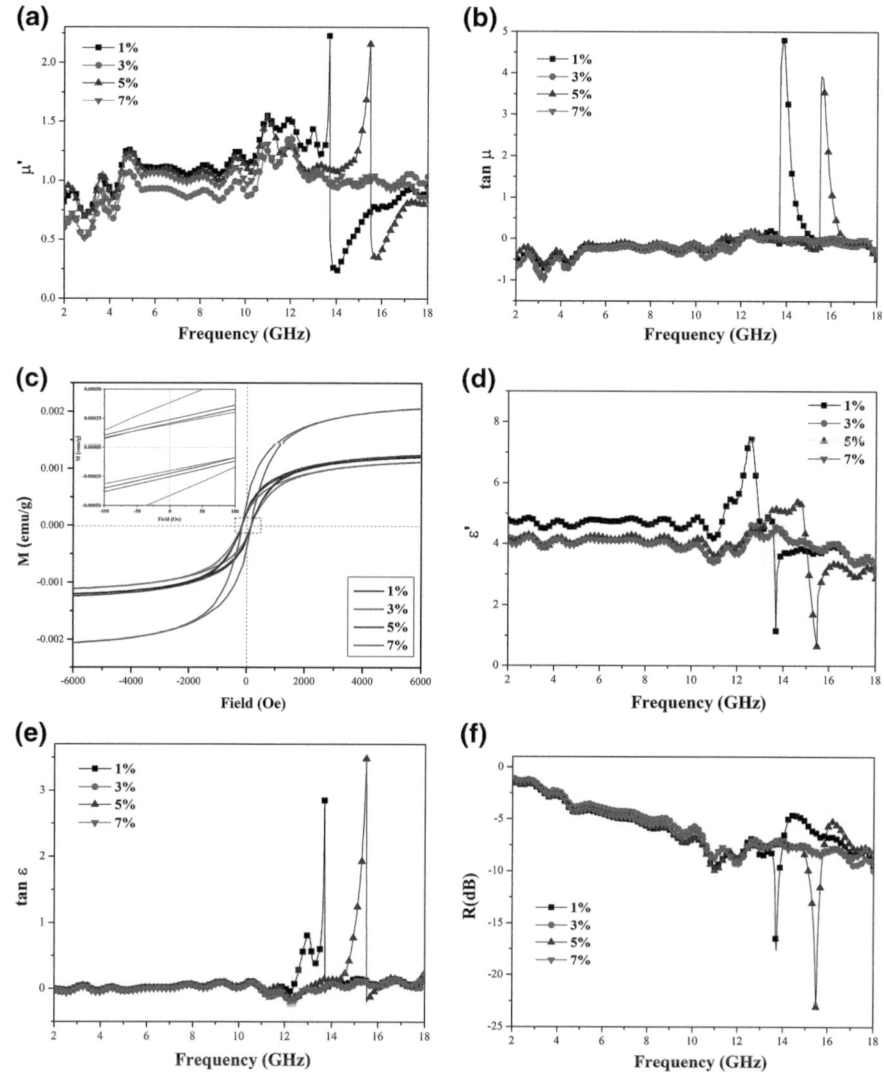

Fig. 4 Electromagnetic properties and wave-absorbing performance of SiCN(Fe)/Si_3N_4 with 1–7 wt% iron(III) acetylacetonate added. **a** μ′; **b** tan μ; **c** magnetic hysteresis loops; **d** ε′; (e) tan ε; **f** reflectivity

indicating that the SiCN composite ceramic is easy to be magnetized and demagnetized, and is a typical soft magnetic material [14].

The relative complex permittivity ($\varepsilon = \varepsilon' - j\varepsilon''$) is the key parameter to characterize the dielectric properties of materials. According to Debye theory, the real part of the relative complex permittivity ε′ is related to polarization relaxation, indicating its polarization ability, while the imaginary part ε″ represents the dielectric loss of the material [15]. The reflectivity R is calculated from the relevant electromagnetic parameters [16]. By observing Fig. 4, it can be seen that the dielectric parameters of the samples are stable at 2–12 GHz, the average value of the real part of the dielectric constant is about 4, the tangent value of the dielectric loss angle floats up and down zero, and the reflectivity is in the range of $0 \sim -10$ dB. While in the range of 12–18 GHz, the parameters of 3 wt% and 7 wt% iron(III) acetylacetone remained stable, and the dielectric parameters of 1 wt% and 5 wt% samples showed natural vibration peaks, which indicated that the internal frequency of the samples was similar to that of external electromagnetic fields, resulting in resonance. When the sample was added with 5 wt% iron(III) acetylacetone, the ε' reached the lowest value of 1 at 15.4 GHz, the tan ε value was the highest peak value of 3.09 at 15.2 GHz, and the R of the sample was lower than −20 dB at 15.5 GHz, which implied that the sample with 5 wt% iron(III) acetylacetone had better wave absorption performance in the high-frequency region and could absorb more than 99% of the incident electromagnetic waves. The difference in dielectric parameters of the samples is mainly related to the phase composition and internal morphology. The content and distribution position of free C, β-SiC decomposed by polymer pyrolysis at high temperature, and reduced α-Fe are of great help to the improvement of wave absorption properties.

Conclusions

SiCN(Fe)/Si_3N_4 ceramics were synthesized by dipping-pyrolysis polysilazane with 1–7 wt% iron(III) acetylacetonate and use porous silicon nitride as matrix. The polysilazane pyrolyzed free carbon and silicon nitride at 1000 °C. The content of iron increased with the increase of the amount of iron(III) acetylacetonate addition. The uniform distribution of C, β-SiC, and α-Fe in the samples is helpful to improve the wave absorption performance of the materials. The ε' reached the lowest value of 1 at 15.4 GHZ, the highest peak value of the tan ε value was the 3.09 at 15.2 GHz, and the R was lower than −20 dB at 15.5 GHz when with 5 wt% iron (III) acetylacetone added. It is indicated that it has great wave-absorbing properties in the high-frequency range.

References

1. Abbas SM, Dixit AK, Chatterjee R et al (2007) Complex permittivity, complex permeability and microwave absorption properties of ferrite-polymer composites. J Magn Magn Mater 309 (1):20–24
2. Tong GX, Ma J, Wu WH et al (2011) Grinding speed dependence of microstructure, conductivity, and microwave electromagnetic and absorbing characteristics of the flaked Fe particles. J Mater Res 26(5):682–688

3. Zhu H, Zhang L (2009) Dielectric, magnetic, and microwave absorbing properties of multi-walled carbon nanotubes filled with Sm_2O_3 nanoparticles. Mater Lett 63(2):272–274
4. Afzalia A, Mottaghitalab V, Afghahi SSS et al (2017) Electromagnetic properties of absorber fabric coated with $BaFe_{12}O_{19}$/MWCNTs/PANi nanocomposite in X and Ku bands frequency. J Magn Magn Mater 442:224–230
5. Wu H, Wang L, Wang Y (2012) Enhanced microwave performance of highly ordered mesoporous carbon coated by Ni_2O_3 nanoparticles. J Alloy Compd 525:82–86
6. Zhang D, Liu Q, Fan T (2008) Electromagnetic wave absorption properties of porous carbon/Co nanocomposites. Appl Phys Lett 93(1):13110–13111
7. Nikmanesh H, Hoghoghifard S, Hadi-Sichani B (2019) Study of the structural, magnetic, and microwave absorption properties of the simultaneous substitution of several cations in the barium hexaferrite structure. J Alloy Compd 775:1101–1108
8. Deng LJ, Han MG (2007) Microwave absorbing performances of multiwalled carbon nanotube composites with negative permeability. Appl Phys Lett 91(2):23111–23119
9. Geng D, Liu X, Xie Z et al (2011) Magnetic and Microwave-absorption properties of Graphite-coated (Fe, Ni) Nanocapsules. J Mater Sci Technol 27(7):607–614
10. Li Q, Yin X, Feng L (2012) Dielectric properties of Si_3N_4–SiCN composite ceramics in X-band. Ceram Int 38(7):6015–6020
11. Liu X, Zhang L, Liu Y et al (2014) Microstructure and the dielectric properties of SiCN–Si_3N_4 ceramics fabricated via LPCVD/CVI. Ceram Int 40(3):5097–5102
12. Lin X, Shi J, Gong H (2017) Preparation and properties of pressureless-sintered porous Si_3N_4 ceramics. Adv Funct Mater (CMC 2017), 3–100
13. Zhang Z, Fan Z, Luo G (2006) Electromagnetic and microwave absorbing properties of multi-walled carbon nanotubes/polymer composites. Mater Sci Eng, B. Solid-State Mater Adv Technol 132(1/2):85–89
14. Liu H, Zou Y, Yang L (2006) The influence of temperature on magnetic and microwave absorption properties of Fe/graphite oxide nanocomposites. J Magn Magn Mater 302(2):343–347
15. Sun Y, Li Y, Su D et al (2015) Preparation and characterization of high temperature SiCN/ZrB_2 ceramic composite. Ceram Int 41(3):3947–3951
16. Liu JR, Itoh M, Machida K (2003) Electromagnetic wave absorption properties of Fe/Fe_3B/Y_2O_3 nanoparticles in gigahertz range. Appl Phys Lett 83:4017

Mechanical Properties of Boron Nitride Nanosheets (BNNSs) Reinforced Si_3N_4 Composites

Guandong Liang, Jianqiang Bi, Guoxun Sun, Yafei Chen and Weili Wang

Abstract The boron nitride nanosheets (BNNSs)/silicon nitride (Si_3N_4) composite ceramics were prepared by SPS sintering. BNNSs with few layers and transverse size in micron-scale were prepared by ball-milling and ultrasound-assisted liquid-phase stripping. In this work, it was found that the BNNSs prepared by these two methods could enhance the mechanical properties of silicon nitride ceramics, and of BNNSs/Si_3N_4 composite ceramics exhibit better bending strength when adding the ultrasound-assisted liquid-phase stripping BNNSs. The bending strength and fracture toughness of BNNSs/Si_3N_4 composite ceramics was increased by 36% and 51%, respectively, when 2 wt% BNNSs prepared by ball-milling method were added. The pullout of BNNSs, the bridging of cracks, and the deflection of cracks are the main factors of improving the toughness.

Keywords Boron nitride nanosheets · BNNSs/Si_3N_4 composite ceramics · Mechanical properties

Introduction

Graphene has been widely used as a reinforcement for ceramic matrix composites due to its excellent mechanical properties [1–3]. BNNSs, commonly known as "white graphene", has similar structural and mechanical properties to graphene. BNNSs also exhibit unique dielectric properties, insulating properties, and high temperature resistance, thus has become a promising enhancer [4].

G. Liang · J. Bi (✉) · G. Sun · Y. Chen · W. Wang
Key Laboratory for Liquid–Solid Structure Evolution and Processing of Materials, Ministry of Education, Shandong University, Jinan 250061, China
e-mail: bjq1969@163.com

G. Liang · J. Bi · G. Sun · Y. Chen · W. Wang
Engineering Ceramics Key Laboratory of Shandong Province, Shandong University, Jinan 250061, China

B. Li et al. (eds.), *Advances in Powder and Ceramic Materials Science*, The Minerals, Metals & Materials Series,
https://doi.org/10.1007/978-3-030-36552-3_8

When graphene is employed as a reinforcing agent, it is easily oxidized by high temperature in the air, which greatly limits its application under high-temperature conditions [5]. In addition, due to its superior electrical conductivity, the application of graphene is also limited in wave-transparent materials and insulating materials. Boron nitride nanosheets have excellent dielectric properties [6] and high temperature resistance [5, 7], thus can replace graphene materials as reinforcing agents for composite materials in this field.

Silicon nitride ceramics have been widely used in high-speed cutting tools, engine parts, and aerospace radome materials due to their high strength, hardness, toughness, excellent wear resistance, thermal shock resistance, low dielectric constant and dielectric loss, and good creep resistance [4, 8–11]. In order to further improve the mechanical properties of the silicon nitride ceramics at high temperature, some reinforcing agents with excellent mechanical properties can be added, and it is obvious that the graphene materials have their limitations. Employing the BNNSs to enhancing the silicon nitride ceramics not only can improve the toughness of the silicon nitride ceramic, but can also reduce the dielectric constant of the silicon nitride ceramic, thus improving the wave transmission performance of the silicon nitride ceramic.

The efficient preparation of boron nitride nanosheets is one of the problems that has been plaguing researchers. Common preparation methods of boron nitride nanosheets include ball-milling, liquid-phase stripping, and chemical-vapor-deposition. The ball-milling method introduces a large number of defects into the nanosheet during the process, but its yield is high and the process is simple [12, 13]; the liquid-phase stripping method has low yield and is not easy to prepare in large quantities, but the prepared boron nitride nanosheets can keep the original crystal structure and have few defects [14–16]; the chemical-vapor-deposition method is expensive, and there are difficulties in that the nanosheet and the substrate are difficult to peel off, but it can realize the controllable preparation of the number of nanosheets [17].

Each of the above three methods for preparing boron nitride nanosheets has its own advantages. As far as we know, there are few reports about silicon nitride composite ceramics containing BNNSs nanosheets. In this paper, boron nitride nanosheets are prepared by ball-milling and liquid-phase stripping, respectively; the effects of nanosheets prepared by two methods on the mechanical properties of silicon nitride composite ceramics were also studied.

Experimental

Preparation of Boron Nitride Nanosheets

Preparation of BNNSs by Ultrasonic-Assisted Liquid-Phase Stripping

A mixed liquid of anhydrous ethanol and deionized water was used as a dispersion. 0.5 g of h-BN powder was added per 1000 ml of the dispersion (55 vol.% of absolute ethanol and 45 vol.% of deionized water [18]), and the mixture was continuously ultrasonicated for 20 h. After 48 h of standing, 40% of the liquid in the upper layer of the mixed disperse solution was taken for filtration, and the powder on the filter membrane was dried at 80 °C for 10 h, and then BNNSs was collected.

Ball-Milling Stripping BNNSs

Weigh 2.5 g of h-BN powder, add 250 ml of absolute ethanol (the ratio of grinding ball to material is 100:1, solvent to material ratio is 80:1) to two ball mill jars with a capacity of 500 ml, according to the ratio. The zirconia grinding ball was added, and after 18 h of ball-milling, a milky white solution with uniform peeling was obtained. The solution was sonicated for 2 h, then filtrated, and dried for 10 h to obtain BNNSs.

Preparation of BNNSs/Silicon Nitride Composite Ceramics

According to the experimental group, BNNSs with different mass fractions were weighed separately (Table 1). After adding a small amount of absolute ethanol, ultrasound was conducted for 2 h. Quantitative powders of silicon nitride, yttrium oxide (6 wt%), and alumina (4 wt%) were obtained according to the proportion. The above substances and a suitable amount of anhydrous ethanol were jointly added to the ball-milling jar. The uniform composite powder was obtained by wet-grinding zirconia ball for 6 h and drying for more than 24 h.

Each group of sufficient mixed powders was added to the graphite mold with a diameter of 30 mm and sintered by SPS. The mixture powders were then heated at a rate of 100 °C/min from room temperature and hot-pressing sintered at 1600 °C for 1 h under the pressure of 50 MPa in a flowing Ar atmosphere before the furnace cooled down to room temperature.

Table 1 Contents of each component in the mixed powder

Group	1	2	3	4	5	6
Mass fraction of BNNSs (%)	0	0.1	0.5	2	0.1	0.5
Mass fraction of Si_3N_4 (%)	90	89.9	89.5	88	89.9	89.5
Mass fraction of Al_2O_3 (%)	4	4	4	4	4	4
Mass fraction of Y_2O_3 (%)	6	6	6	6	6	6

Note BNNSs doped in No. 2, No. 3, and No. 4 are obtained by ball-milling method, while BNNSs doped in No. 5 and No. 6 are obtained by ultrasound-assisted peeling method

Tests and Characterization

The bending strength of the bar specimens (3.0 mm × 4.0 mm × 25 mm) was measured by a three-point bending test with a span of 20 mm at a crosshead speed of 0.5 mm/min. Three specimens were tested in each group. Besides, fracture toughness of three specimens (2.0 mm × 4.0 mm × 25 mm) for each sample was measured by the single-edge notched beam (SENB) method. A notch of 2.0 mm in depth and 0.3 mm in width was introduced in the middle of each specimen by a thin diamond blade. Both the bending strength and fracture toughness tests were conducted on a CMT5105 electromechanical universal testing machine (Shenzhen SANS Testing Machine Co., Ltd., Shenzhen, Guangdong, China). In addition, the morphology of BNNSs and the fracture surfaces of the composites were examined via thermal field emission scanning electron microscope (FESEM; Hitachi SU-70, Tokyo, Japan).

Results and Discussion

Figure 1 shows the SEM images of BNNSs prepared by two methods. From Fig. 1a, b, it can be seen that many fragments are adsorbed on the surface of boron nitride nanosheets prepared by ball-milling. One possible reason is that some nanosheets are grounded into fragments by high-speed rotating grinding balls, and these fragments are easily adsorbed on the surface of large-scale nanosheets. In contrast, few fragments were found in boron nitride nanosheets prepared by ultrasound-assisted liquid-phase peeling in Fig. 1c, d.

By comparing the two methods, it can be found that the nanosheets prepared by ultrasonic peeling have larger transverse size, but the nanosheets prepared by ball-milling are thinner. There are some transparent and edge-curled nanosheets in the nanosheets prepared by the two methods, and the transverse size is in the micron level. It shows that both methods can produce a few layers and large sheets of boron nitride nanosheets [5], but the quality of the nanosheets prepared by ultrasound-assisted liquid-phase peeling is better.

Fig. 1 FESEM images of BNNSs. **a** and **b** are BNNSs prepared by ball-milling, **c** and **d** are BNNSs prepared by ultrasonic-assisted liquid-phase stripping

Figure 2 showse the fracture scanning images of BNNSs/Si_3N_4 composites. Figure 2a–f shows that most of the composites are rod-like β-phase grains, which indicates that the conversion of silicon nitride powder from α-phase to β-phase is sufficient during sintering. These different orientations of the β-phase grains have a positive impact on the mechanical properties of silicon nitride ceramics [19]. The presence of boron nitride nanosheets can be observed in Fig. 2c, d, f, which can further improve the mechanical properties of silicon nitride composite ceramics. In Fig. 2b, e, no boron nitride nanosheets were found, possibly due to the small amount of added nanosheets.

Figure 3 shows a histogram of mechanical properties of Si_3N_4 composite ceramics prepared by adding ultrasound-assisted liquid-phase peeling BNNSs. The results show that the bending strength and fracture toughness of the composite ceramics increase to a certain extent with the increase in the amount of BNNSs prepared by ultrasonic-assisted liquid-phase peeling. The bending strength of BNNSs/Si_3N_4 composite ceramics increased from 402.0 MPa to 482.7 MPa, and the fracture toughness increased from 3.61 MPa·$m^{1/2}$ to 4.16 MPa·$m^{1/2}$, after adding 0.5 wt% BNNSs nanosheets prepared by ultrasonic-assisted liquid-phase peeling. The bending strength and fracture toughness of boron nitride nanosheets prepared by 0.1 wt% ultrasound-assisted liquid-phase peeling increased significantly, which may be related to the measurement errors caused by many notches on the surface of the sample during processing.

Figure 4 shows a histogram of mechanical properties of silicon nitride composites prepared by adding boron nitride nanosheets obtained by ball-milling. Similarly, the flexural strength and fracture toughness of the composite ceramics

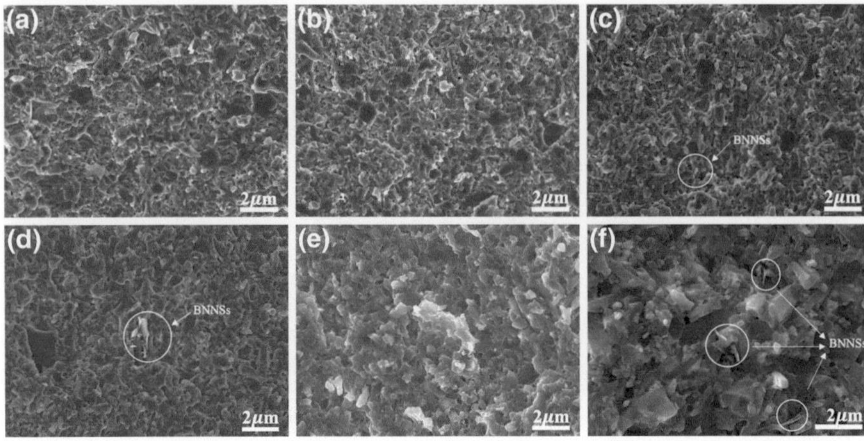

Fig. 2 FESEM images of the fracture surfaces of BNNSs/Si_3N_4 composites: **a**–**f** correspond to samples no. 1–6 in Table 1, respectively

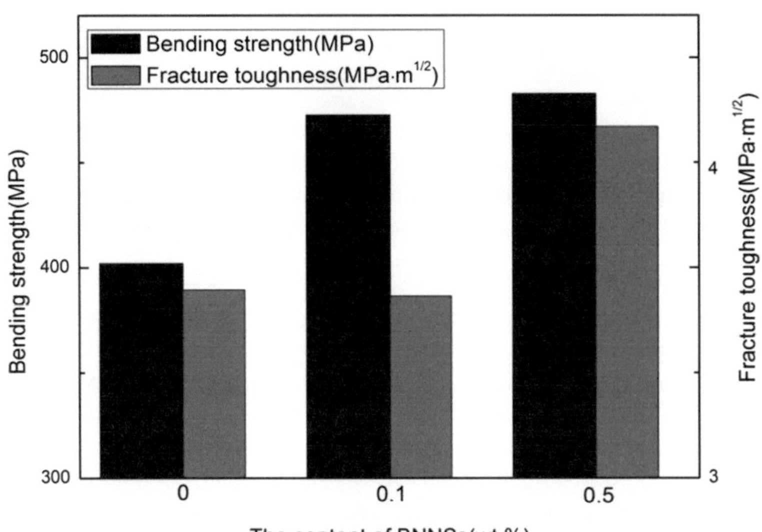

Fig. 3 Mechanical properties of BNNSs/Si_3N_4 composite ceramics prepared by ultrasonic-assisted liquid-phase stripping BNNSs

increased with the addition of nanosheets; the flexural strength and fracture toughness of the composite ceramics reached the maximum, 547.0 MPa and 5.47 MPa·$m^{1/2}$, respectively, when adding 2 wt% BNNSs obtained. Compared with the Si_3N_4 ceramics without boron nitride nanosheets, the flexural strength and fracture toughness are 36% and 51% higher, respectively.

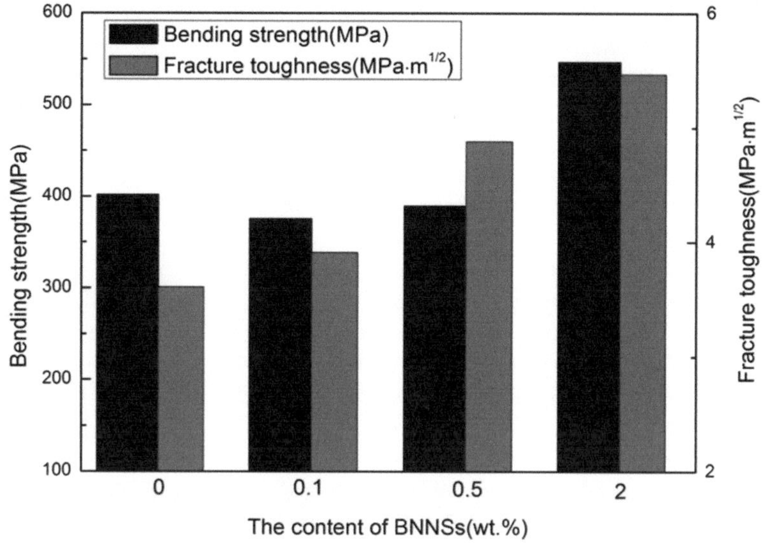

Fig. 4 Mechanical properties of BNNSs/Si_3N_4 composite ceramics prepared by ball-milling BNNSs

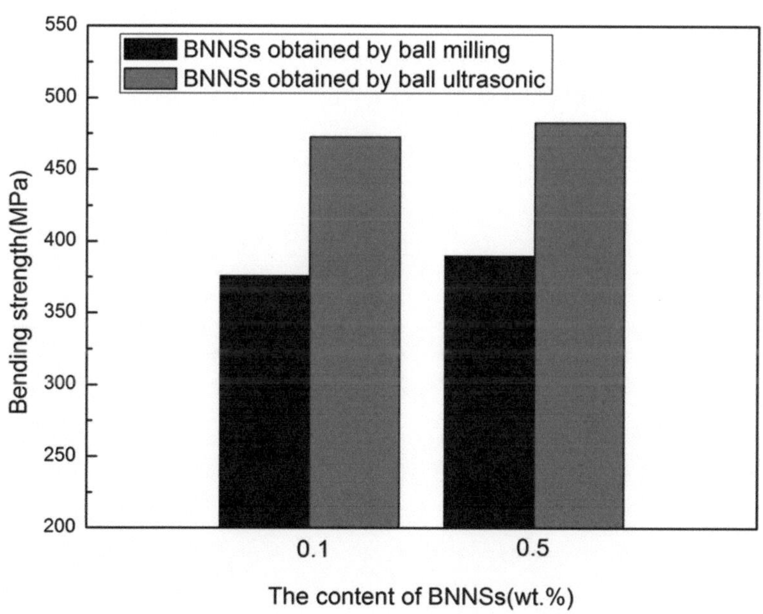

Fig. 5 Comparison of bending strength of BNNSs/Si_3N_4 composite ceramics after adding BNNSs prepared by different methods

Figure 5 shows a comparison of the bending strength of composite ceramics after adding the same mass fraction of boron nitride nanosheets prepared by two methods. It can be seen from the figure that when the same mass fraction of boron nitride nanosheets is added, the bending strength of silicon nitride ceramics can be enhanced more when the ultrasound-assisted liquid-phase stripping BNNSs are added. This is because boron nitride nanosheets obtained by ultrasound-assisted liquid-phase stripping method have fewer defects and more complete crystal structure than those obtained by ball-milling method.

The toughening mechanisms were clearly identified from the FESEM images of the indentation crack propagation paths and the fracture surfaces. FESEM images of Vickers indentation imprints performed on the sample surface are presented in Fig. 6. The pulling mechanism of boron nitride nanosheets can be clearly observed from Fig. 6c, d, and some nanosheets remain in the ceramic matrix [5, 20]. From the crack growth path of indentation, the deflection of the crack and the bridging phenomenon of the boron nitride nanosheets can be clearly observed, and the crack growth path is serrated. These will consume a lot of fracture energy, which is conducive to the improvement of fracture toughness of composite ceramics [21].

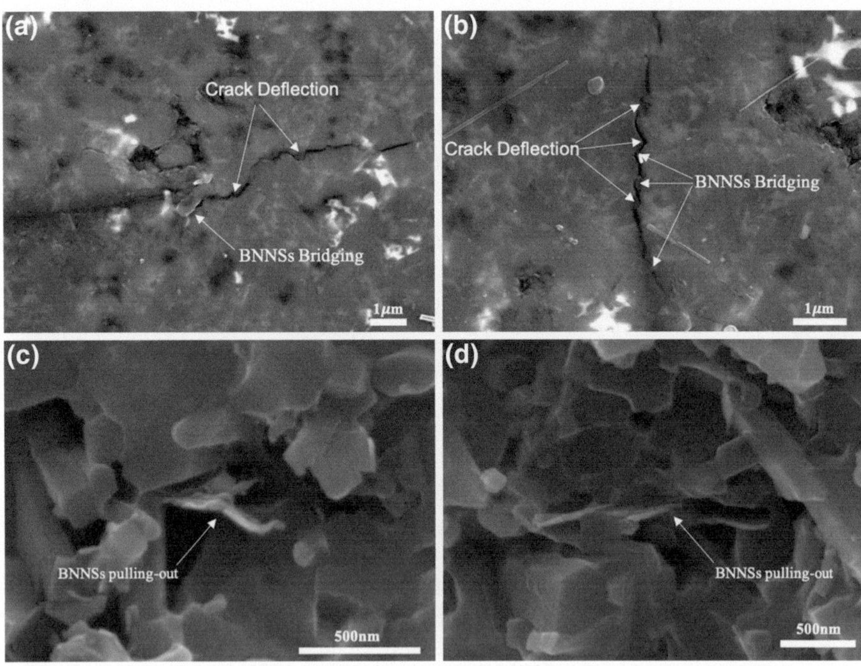

Fig. 6 Toughening mechanisms. **a** and **b** details of indentation crack propagation with crack deflection and BNNSs bridging in sample 6. **c** and **d** FESEM images of the fracture surface of sample 6 after three-point bending tests

Conclusion

BNNSs with few layers and transverse size in micron-scale were prepared by ball-milling and ultrasound-assisted liquid-phase stripping. We found that the BNNSs prepared by two methods could enhance the mechanical properties of silicon nitride ceramics, and the silicon nitride ceramics show better mechanical properties with the addition of ultrasound-assisted liquid-phase stripping BNNSs. The bending strength and fracture toughness of silicon nitride composite ceramics were increased by 36% and 51%, respectively, when BNNSs prepared by ball-milling method of 2 wt% were added. The pullout of BNNSs, the bridging of cracks, and the deflection of cracks are the main factors to improve the toughness.

References

1. Rafiee MA, Rafiee J, Wang Z et al (2009) Enhanced mechanical properties of nanocomposites at low graphene content. ACS Nano 3(12):3884
2. Liu Y, Wu H, Chen G (2016) Enhanced mechanical properties of nanocomposites at low graphene content based on in situ ball milling. Polym Compos 37(4):1190–1197
3. Wang K, Wang Y, Fan Z, Yan J, Wei T (2011) Preparation of graphene nanosheet/alumina composites by spark plasma sintering. Mater Res Bull 46:315–318
4. Lee B, Lee D, Lee JH et al (2016) Enhancement of toughness and wear resistance in boron nitride nanoplatelet (BNNP) reinforced Si_3N_4 nanocomposites. Sci Rep 6:27609
5. Sun G, Bi J, Wang W et al (2017) Microstructure and mechanical properties of boron nitride nanosheets-reinforced fused silica composites. J Eur Ceram Soc 37(9):3195–3202
6. Shi G, Hanlumyuang Y, Liu Z et al (2014) Boron Nitride-Graphene Nanocapacitor and the origins of anomalous size-dependent increase of capacitance. Nano Lett 14(4):1739–1744
7. Li LH, Cervenka J, Watanabe K, Taniguchi T, Chen Y (2014) Strong oxidation resistance of atomically thin boron nitride nanosheets. ACS Nano 8:1457–1462
8. Riley FL (2000) Silicon nitride and related materials. J Am Ceram Soc 83:245–265
9. Bocanegra-Bernal MH, Matovic B (2010) Mechanical properties of silicon nitride-based ceramics and its use in structural applications at high temperatures. Mater Sci Eng A 527:1314–1338
10. Wang Z, Jia J, Cao L et al (2019) Microstructure and mechanical properties of spark plasma sintered Si3N4/WC ceramic tools. Materials 12(11):1868
11. Miranzo P, Jesús González-Julián, María Isabel Osendi et al (2011) Enhanced particle rearrangement during liquid phase spark plasma sintering of silicon nitride-based ceramics. Ceram Int 37(1):159–166
12. Lei W, Liu D, Chen Y (2015) Highly crumpled boron nitride nanosheets as adsorbents: scalable solvent-less production. Adv Mater Interfaces 2:1400529–1400534
13. Li LH, Chen Y, Behan G et al (2011) Large-scale mechanical peeling of boron nitride nanosheets by low-energy ball Milling. J Mater Chem 21:11862–11867
14. Ma P, Spencer JT (2014) Non-covalent stabilization and functionalization of boron nitride nanosheets (BNNSs) by organic polymers: formation of complex BNNSs-containing structures. J Mater Sci 50:313–323
15. Gao G, Gao W, Cannuccia E et al (2012) Artificially stacked atomic layers: toward new van der waals solids. Nano Lett 12:3518–3525

16. Zhi C, Bando Y, Tang C et al (2009) Large scale fabrication of boron nitride nanosheets and their utilization in polymeric composites with improved thermal and mechanical properties. Adv Mater 21:2889–2893
17. Muller F, Hufner S, Sachdev H et al (2010) Epitaxial growth of hexagonal boron nitride on Ag (111). Phys Rev B 82:113406–113409
18. Zhou KG, Mao NN, Wang HX, Peng Y, Zhang HL (2011) A mixed-solvent strategy for efficient exfoliation of inorganic graphene analogues. Angew Chem Int Ed 50:10839–10842
19. Kumar A, Gokhale A, Ghosh S et al (2019) Effect of nano-sized sintering additives on microstructure and mechanical properties of Si_3N_4 ceramics. Mater Sci Eng A 750:132–140
20. Chen YF, Bi JQ, Wang WL et al (2014) Toughening in boron nitride nanotubes/silicon nitride composites. Mater Sci Eng A 590:16–20
21. Chen C, Pan L, Li X et al (2017) Mechanical and thermal properties of graphene nanosheets/ magnesia composites. Ceram Int 43(13):10377–10385

Part V
Poster Session

Polymer Fibers from Waste Tires and Sugarcane Molasses for Soil Improving

Juan Esteban Jimenez Hoyos and Henry A. Colorado L

Abstract This research shows results about the use of recycled tire polymer fibers (RTPF) and sugarcane molasses, both blended with kaolin clay for soil improving applications. RTPF were obtained from a tire recycled company and the molasses was obtained from a sugarcane manufacturer. Both materials are considered as wastes in some locations and therefore the first positive impact of this research is to have a green solution for these by-products. The materials characterization was performed with scanning electron microscopy and compression strength tests. Results showed that unconfined compressive strength improved from about 1419 kPa for the sample without the addition of fibers and molasses to 2037 kPa for the sample with 0.1 wt% of fibers and 2.0 wt% molasses, contents taken with respect to the dry weight of the soil.

Keywords Waste tires · Polymer fibers · Molasses · Clay · Soil improving · Soil microorganisms

Introduction

The accumulation of tires that ended their useful life has become a worldwide problem that has been the subject of study by the academic community for years, but due to the scale of the car's demand, the problem is still far from a solution. Therefore, innovative reuse methods are required in order to contribute to solve the problem via initiatives such as circular economy, particularly in developing regions such as Latin America [1–3], where multiples options are now in developing such

---

J. E. J. Hoyos · H. A. Colorado L (✉)
CCComposites Laboratory, Universidad de Antioquia UdeA,
Calle 70 N°. 52-21, Medellín, Colombia
e-mail: henry.colorado@udea.edu.co

J. E. J. Hoyos · H. A. Colorado L
Facultad de Ingeniería, Universidad de Antioquia, bloque 20, 67 St. #. 53 - 108, Medellín, Antioquia, Colombia

B. Li et al. (eds.), *Advances in Powder and Ceramic Materials Science*,
The Minerals, Metals & Materials Series,
https://doi.org/10.1007/978-3-030-36552-3_9

as the tire rubber waste addition to pavements [4] and to cementitious materials [5]. Every year, 1.4 trillion tires are produced worldwide, which generate 17 million tons of used tires per year [6]. Only in 2017, 4.7 million tons of tires were discarded in the United States [7].

To supply the high mechanical stresses in the tires, these are made of rubber reinforced with steel and polymer textile. The compound generated by these materials favors the conditions so that the tires have long degradation times. Cross-links between the rubber polymer chains, additives, and stabilizers used in tires manufacturing, make the tires resistant to biodegradation and even to hard environmental conditions [8].

Sienkiewicz et al. [6] identified five processes that can be used for the handling of discarded tires: retreading, energy recovery, pyrolysis, product recycling, and material recycling. Retreading is the way in which the worn tread is replaced with a new one [9], although, some results show that vehicle safety is affected by retreaded tires [10]. Energy recovery is the way of tire recovering, where the tires are used as an energy raw material [1]: used tires have a high calorific value of about 32 MJ/kg [11]. Pyrolysis is the process that converts waste tires to flammable gas, pyrolytic oil, carbon black liquid fuel, and pyrolysis char [12, 13]. Product recycling means to recycle of entire used tires in their original form [1]. Material recycling is realized as mechanical grinding of tires, which yields rubber materials of different degrees of grinding [14].

Multiple research has shown results regarding used tire-recycling by-products in engineering problems. This is the case of RTPF and recycled rubber from discarded tires, which were included in different types of concrete to assess the change in mechanical properties [15–20]. Steel fibers have been studied as a reinforcing material for concrete [21, 22]. And rubber particles of different sizes and shapes from the tire-recycling process have been mixed with clay and sandy soils to improve engineering properties [23–26].

Molasses is an organic by-product of sugar, dark brown in color, highly viscous, and with a strong odor. Its main components are carbon and oxygen; while magnesium, sulfur, chlorine, potassium, and calcium are minor elements of its composition.

In the field of engineering, molasses has been used in different areas. It has been used to tailor properties of concrete and cement [27–29], as a stabilizer of expansive clays [30], and has been also investigated as a potential replacement of conventional materials for roads infrastructure [31].

Kaolin is a white clay soil from residual or sedimentary deposits, mainly composed of silica and aluminum. Its structural formation is constituted by octahedral aluminum and tetrahedral silica sheets [32]. The kaolin has been used in several investigations related to construction materials such as calcined kaolin, included in concrete mixes for better durability and mechanical strength [33–35].

Soil improvement is an engineering branch that is responsible for developing methods and materials to modify the engineering properties of the soil and thus converting it into an adequate space to develop construction projects. Mitchel [36] performed a comprehensive state of the art and classified different methods for soil improvement; later Terashi and Juran [37], based on Mitchel's work, reclassified the methods by fitting them into eight groups: replacement, densification, consolidation, grouting, admixtures stabilization, thermal stabilization, reinforcement, and miscellaneous.

In order to contribute to the state of the art in the management of waste from different industrial and agricultural processes, an environmentally friendly alternative for the disposal of two wastes was studied in this work. Molasses was used as a bonding material between soil particles, and matrix for fiber dispersion. Characterization tests were performed with scanning electron microscopy, and the soil improvement was evaluated via UCS tests.

Experimental Procedure

Materials

A tire recycled company located at Rionegro, Colombia, provided the waste fibers. This polymer fibers have an average of 22.4 μm in diameter, and came mixed with residual rubber. The residual rubber is average 46.0 μm and its concentration is about 60% by fiber–rubber weight. Molasses is a by-product of sugar manufacturing and it was supplied by a sugarcane manufacturer located at Medellin, Colombia. Kaolin soil in natural conditions used in this investigation was extracted from a regional residual quarry at Sonsón, Antioquia.

Soil Samples

Kaolin soil was crushed using hand roller until a homogenous grain-size was achieved. Molasses, polymer fibers, and water contents were all mixed using a Hobart N50 mixer machine for 5 min, with molasses and fibers previously mixed by hand for 3 min. The mixed soil was compacted according to the ASTM D1557–02 method A. Three soil samples were taken by each compacted mixture. 2" diameter aluminum tubes were introduced in mold with the help of a hammer. Then, the soil sample was extracted from the aluminum tube using a hydraulic jack. Then, the soil samples were covered with a plastic wrap. Finally, samples were kept in a room at 20 ± 3 °C and air atmosphere for curing for 28 days.

Tests

Soil improving was assessed conducting Unconfined Compression Strength test (UCS) according to ASTM D2166, which was performed using a Shimadzu AG250 KN universal testing machine at 0.68 mm/s. Dehydrated and gold sputtering soil samples were analyzed with a scanning electron microscopy (SEM) apparatus, a JEOL JSM 6700 R in high vacuum mode. Soil samples were dehydrated putting it in furnace at 30°C for 24 h. Gold sputtering was conducted using a Hummer 6.2 equipment at 15 mA AC for 30 s. Based on the SEM and optical images, the fiber diameter and the residual rubber grain-size were estimated using Image J software.

Results and Analysis

Table 1 shows the material composition of each soil mixture. Molasses and fibers content were calculated in the dry soil. Soil was mixed with water until 35% moisture. Seven types of samples were made, the fiber content was constant while the molasses content varied from 0% to 12% based on the dry weight of the soil.

Figure 1 shows rthe aw materials used in this research. Figure 1a represents a deposit of used tires, which are usually the final disposal sites for used tires. Figure 2b represents polymer fibers on micro-scale. The fibers are connected one to another due to recycling process, which destroys the polymer textile and forms short fibers. Figure 2c presents the real color and viscosity of molasses. Figure 2d is kaolin soil in natural conditions, with different particles size.

Figure 2 shows the diameter distribution of the fibers. These fibers have a diameter ranging from 10 to 40 μm, and with 22.4 μm as average diameter.

Figure 3 summarizes the residual rubber grain-size probabilistic distribution. Residual rubber grain can be from 15 μm to 105 μm in diameter; with 46.03 μm of average diameter.

Table 1 Soil mixtures

Designation	Molasses (%)	Fibers (%)	Soil (%)	Total % (By dry soil)	% Water content
M0	0	0	100	100	35
M2	2	0.1	97.9	100	35
M4	4	0.1	95.9	100	35
M6	6	0.1	93.9	100	35
M8	8	0.1	91.9	100	35
M10	10	0.1	89.9	100	35
M12	12	0.1	87.9	100	35

Fig. 1 Materials **a** deposit of used tires **b** waste tire fibers, **c** molasses, and **d** kaolin soil in natural conditions

Fig. 2 Nylon fibers diameter probability distribution

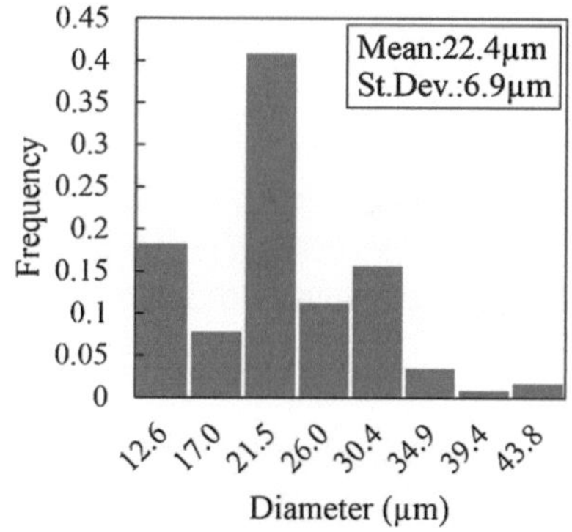

Fig. 3 Residual rubber grain-size probability distribution

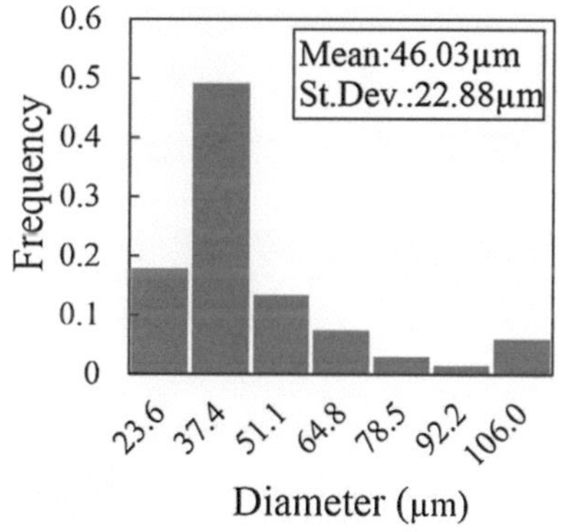

Figure 4 exposes UCS test results. The highest soil strength was 2037 kPa, which was achieved when 2% molasses content by dry soil was added. It was seen that as the molasses content increases, the ductility of the soil specimen increases as well.

Fig. 4 Typical UCS test

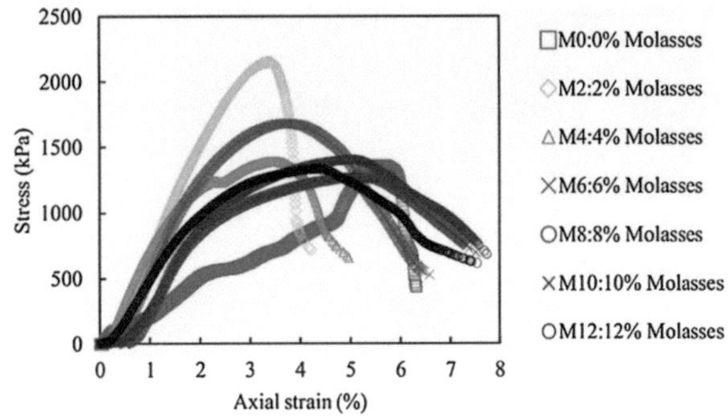

Figure 5 shows SEM images. Figure 5a is a 2000X SEM image which shows the end of a single fiber. Figure 5b represents M2 soil sample at 2000X, which demotes the fibers and soil interface. Figure 5c exposes M4 soil sample at 500X, which shows the random dispersion of the fibers in the soil matrix. Figure 5d corresponds to M0 soil sample 500X. Figure 5e shows M8 soil sample at 500x, which shows the fiber coming out from the soil matrix. Figure 5f presents M12 soil sample at 500X, where the adhesion between the fibers and the soil with high molasses contents is observed.

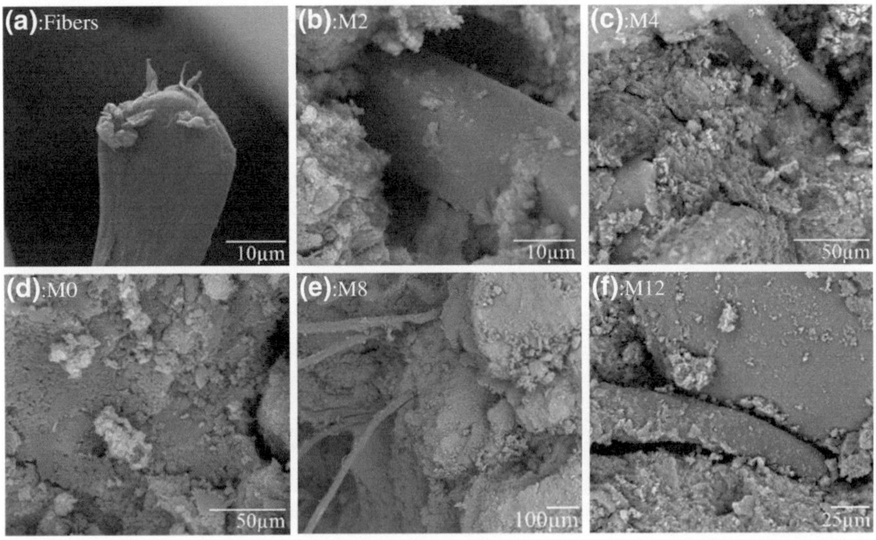

Fig. 5 SEM images for the soil composite fabricated in this research

Discussion

To some extent, molasses can be an unconventional material with cementing properties that could be used for soil improvement. Molasses is an organic product, soluble in water, and without toxic components that may pose a threat to tributaries, ecosystems, and people. It is a by-product, cheap and easy to acquire. Only low dosages are necessary to reach the highest strength, which corresponds to an approximate 20 kg of molasses being necessary to treat a ton of soil, and in this way, the maximum soil resistance would be achieved. The research done by Abdulhussein Saeed et al. [38] shows that a soil with similar characteristics and similar laboratory conditions used in this investigation required 10% cement to reach a resistance of 2100 kPa. In this context, for soil improvement under dry conditions, molasses can be a substitute for cement, and 2% molasses would roughly equal 10% cement. Finally, this research has shown the potential of the soil composite developed here, which upon optimization can be used in more applications, perhaps 3D printed for complex shapes with tailored compositions as other clay-based materials [39, 40].

Conclusions

- RTPF and molasses blend is an environment-friendly, economical, and technical option for soil improving. Molasses is ecological by-product without toxic elements which is able to be applied without causing damage to environment. Molasses viscosity permits to separate randomly RTPF in soil ground and generates randomly distribute composite.
- An improvement of about 43% was achieved, going from 1419 kPa for the base line sample to 2037 kPa for the mixture with 2% of molasses and 0.1% of RTPF by dry weight of the soil.

References

1. Ordóñez-Ordóñez E, Echeverry-Lopera G, Colorado-Lopera H (2019) Engineering and economics of the hazardous wastes in Colombia: the need for a circular economy model. Informador Técnico 83(2):155–173
2. Lopera DC, Lopera GIE, Lopera HC (2019) Logistics as an essential area for the development of the solid waste management in Colombia. Informador técnico 83(2):131–154
3. Rúa-Restrepo JJ, Echeverri GI, Colorado HA (2019) Toward a solid waste economy in Colombia: An analysis with respect to other leading economies and Latin America. In: REWAS 2019 (pp 337–354). Springer, Cham
4. Agudelo G, Cifuentes S, Colorado HA (2019) Ground tire rubber and bitumen with wax and its application in a real highway. J Clean Prod 228:1048–1061

5. Revelo CF, Colorado HA (2019) Waste tire rubber in calcium phosphate cement blends. Adv Ceram Environ Funct Struct Energy Appl II Ceram Trans 266:237
6. Sienkiewicz M, Kucinska-Lipka J, Janik H, Balas A (2012) Progress in used tyres management in the European Union: a review. Waste Manag 32(10):1742–1751
7. Rubber Manufactures Association (2018) 2017 US scrap tire management summary. Washington
8. Stevenson K, Stallwood B, Hart AG (2008) Tire rubber recycling and bioremediation: a review. Bioremediat J 12(1):1–11
9. Lebreton B, Tuma A (2006) A quantitative approach to assessing the profitability of car and truck tire remanufacturing. Int J Prod Econ 104(2):639–652
10. Zebala J, Ciepka P, Reza A, Janczur R (2007) Influence of rubber compound and tread pattern of retreaded tyres on vehicle active safety. Forensic Sci Int 167(2–3):173–180
11. Gieré R, Smith K, Blackford M (2006) Chemical composition of fuels and emissions from a coal + tire combustion experiment in a power station. Fuel 85(16):2278–2285
12. Zhang X, Wang T, Ma L, Chang J (2008) Vacuum pyrolysis of waste tires with basic additives. Waste Manag 28(11):2301–2310
13. Choi GG, Jung SH, Oh SJ, Kim JS (2014) Total utilization of waste tire rubber through pyrolysis to obtain oils and CO_2 activation of pyrolysis char. Fuel Process Technol 123:57–64
14. DeSK, Isayev A, Khait K (Eds) (2005) Rubber recycling. CRC Press
15. Baricevic A, Pezer M, Rukavina MJ, Serdar M, Stirmer N (2018) Effect of polymer fibers recycled from waste tires on properties of wet-sprayed concrete. Constr Build Mater 176:135–144
16. Serdar M, Baričević A, Jelčić Rukavina M, Pezer M, Bjegović D, Štirmer N (2015) Shrinkage behaviour of fibre reinforced concrete with recycled tyre polymer fibres. Int J Polymer Sci
17. Baričević A, Rukavina MJ, Pezer M, Štirmer N (2018) Influence of recycled tire polymer fibers on concrete properties. Cement Concr Compos 91:29–41
18. Bignozzi MC, Sandrolini F (2006) Tyre rubber waste recycling in self-compacting concrete. Cem Concr Res 36(4):735–739
19. Huang B, Li G, Pang SS, Eggers J (2004) Investigation into waste tire rubber-filled concrete. J Mater Civ Eng 16(3):187–194
20. Son KS, Hajirasouliha I, Pilakoutas K (2011) Strength and deformability of waste tyre rubber-filled reinforced concrete columns. Constr Build Mater 25(1):218–226
21. Papakonstantinou CG, Tobolski MJ (2006) Use of waste tire steel beads in Portland cement concrete. Cem Concr Res 36(9):1686–1691
22. Pilakoutas K, Neocleous K, Tlemat H (2004) Reuse of tyre steel fibres as concrete reinforcement. Proc ICE-Eng Sustain 157(3):131–138
23. Hataf N, Rahimi MM (2006) Experimental investigation of bearing capacity of sand reinforced with randomly distributed tire shreds. Constr Build Mater 20(10):910–916
24. Jafari M, Esna-ashari M (2012) Effect of waste tire cord reinforcement on unconfined compressive strength of lime stabilized clayey soil under freeze–thaw condition. Cold Reg Sci Technol 82:21–29
25. Yoon YW, Heo SB, Kim KS (2008) Geotechnical performance of waste tires for soil reinforcement from chamber tests. Geotext Geomembr 26(1):100–107
26. Srivastava A, Pandey S, Rana J (2014) Use of shredded tyre waste in improving the geotechnical properties of expansive black cotton soil. Geomech Geoeng 9(4):303–311
27. Aalm A, SinghP (2016) Experimental study on strength characteristics of cement concrete by adding sugar waste. Int J Enhanc Res Sci Technol Eng ISSN, 2319–7463
28. Akar C, Canbaz M (2016) Effect of molasses as an admixture on concrete durability. J Clean Prod 112:2374–2380
29. Ismail N, Hamidah MS, Jelani H, Mansor H, Faizal MM (2014) The studies on the effect of molasses in effective microbed cement paste. In: International Conference on Agriculture, Environment and Biological Science
30. M'Ndegwa JK (2011) The effect of cane molasses on strength of expansive clay soil. J Emerg Trends Eng Appl Sci 2(6):1034–1041

31. Shirsavkar SS, Koranne S (2010) Innovation in road construction using natural polymer. Electron J Geotech Eng 15(1):1614–1624
32. Prasada MS, Reid KJ, Murray HH (1991) Kaolin: processing, properties and application. Appl Clay Sci 6(2):87–119
33. Wong HS, Razak HA (2005) Efficiency of calcined kaolin and silica fume as cement replacement material for strength performance. Cem Concr Res 35(4):696–702
34. Vu DD, Stroeven P, Bui VB (2001) Strength and durability aspects of calcined kaolin-blended Portland cement mortar and concrete. Cement Concr Compos 23(6):471–478
35. Shafiq N, Nuruddin MF, Khan SU, Ayub T (2015) Calcined kaolin as cement replacing material and its use in high strength concrete. Constr Build Mater 81:313–323
36. Mitchell JK (1981) Soil improvement-state of the art report. In: Proceedings 11th international conference on SMFE (Vol 4, pp 509–565)
37. Terashi M, Juran I (2000, November) Ground Improvement-state of the art. In: ISRM international symposium international society for rock mechanics and rock engineering
38. Abdulhussein Saeed K, Anuar Kassim K, Nur H (2014) Physicochemical characterization of cement treated kaolin clay. Građevinar 66(06):513–521
39. Ordoñez E, Gallego JM, Colorado HA (2019) 3D printing via the direct ink writing technique of ceramic pastes from typical formulations used in traditional ceramics industry. Appl Clay Sci 182:105285
40. Revelo CF, Colorado HA (2018) 3D printing of kaolinite clay ceramics using the Direct Ink Writing (DIW) technique. Ceram Int 44(5):5673–5682

Evaluation of the Incorporation of Marble and Granite Residue in Coating Mortars

Euzebio Zanelato, Jonas Alexandre, Afonso Azevedo, Markssuel Marvila, Gustavo Xavier and Sergio Monteiro

Abstract The disposal of ornamental rock waste generates serious environmental impacts, where incorporation in construction materials would be a solution with low environmental impact. The objective of this work is to evaluate the performance of the incorporation of waste from the industry of both marble and granite in mortars. The residues were incorporated at different levels of incorporation (25, 50, 75 and 100%) using different mortar traces (1:1:6 and 1:2:9). The materials were characterized and the mortars were evaluated in the fresh state by the consistency index tests and Squeeze Flow. In the hardened state, mechanical strength, water absorption, and tensile strength tests were performed. The results indicated that the marble residue presents great potential of use, while the incorporation of granite shows a loss of performance.

Keywords Granite · Marble · Mortar

E. Zanelato (✉) · J. Alexandre · M. Marvila · G. Xavier
LECIV – Civil Engineering Laboratory, UENF State University of the Northern Rio De Janeiro, Av. Alberto Lamego, 2000, Campos dos Goytacazes, Rio De Janeiro 28013-602, Brazil
e-mail: ebzanelato@gmail.com

E. Zanelato · A. Azevedo
IFF - Federal Institute Fluminese, DIRINF – Directorate of Infrastructure Rectory, Rua Cel. Valter Kramer, 357 - Parque Vera Cruz, Campos dos Goytacazes, Rio De Janeiro 28080-565, Brazil

A. Azevedo
TER – Department of Agricultural Engineering and Environment, UFF - Federal Fluminense University, Rua Passo da Pátria, 341, Niterói, Rio de Janeiro 24210240, Brazil

S. Monteiro
Department of Materials Science, IME - Military Institute of Engineering, Square General Tibúrcio, 80, Rio de Janeiro 22290-270, Brazil

B. Li et al. (eds.), *Advances in Powder and Ceramic Materials Science*, The Minerals, Metals & Materials Series,
https://doi.org/10.1007/978-3-030-36552-3_10

Introduction

The Brazilian ornamental stone industry presents excellent production and export performance with each passing year. The ornamental stone sector has an important share of the market, where in 2017 alone totaled US$ 1,107.1 million and 2.36 million tons in exports. Among the main responsible for the Brazilian ornamental stone market, Cachoeiro de Itapemirim—ES stands out as a pole of approximately 600 companies [1].

The processing performed on the rock to add value to the raw material is composed of cutting and polishing [2]. Both steps generate a huge amount of waste, which causes great environmental impacts in the region.

The production process of these ornamental rocks produces about 800,000 tons per year of waste in the states of Espirito Santo, Minas Gerais, Bahia, and Ceará. In general, waste is discharged into the environment without prior treatment, thus the industrial sector is penalized by environmentalists for damage to the local environment. Even places where this waste is deposited and collected suffer from the cost and lack of space in landfills [3].

An alternative to lessen the impact caused by this waste is to incorporate it into segments of the building industry, such as mortar and ceramic tiles. Waste reuse would decrease or even eliminate the volume of waste disposed of in landfills

The incorporation of the residue has already shown satisfactory results in incorporations in the red ceramic pieces, where there was an increase in mechanical performance and durability [4].

Mortar incorporation has also been tested by several authors [5–7], which also presented satisfactory results regarding rheology and also mechanical performance.

The objective of this work is to contribute to the increase of research related to the incorporation of residues from the ornamental rock industry and, more specifically, to compare the performance of mortars made of granite and marble in different levels of incorporation (0, 25, 50, 75, and 100%) using two different mortar strokes (1:1:6 and 1:2:9).

Materials and Methods

The materials used to make the mortar were characterized by grain size and grain density tests.

Two traces of mortar were used: 1:1:6 and 1:2:9 (Cement: Hydrated Lime: Sand). These traces were chosen because of their recurring use and proper plasticity for coating application. The 1:2:9 trace exhibits greater water retention and is widely used in locations that are most vulnerable to water loss by evaporation or absorption of the substrate where it was applied. The 1:1:6 trait exhibits greater mechanical strength as well as greater adhesion potential, however, depends on

Table 1 Tests performed on mortar and respective standards

Test	Standard
Consistency index	NBR 13276 [8]
Compressive strength	NBR 13279 [9]
Flexural tensile strength	NBR 13279 [9]
Water absorption	NBR 9778 [10]
Tensile strength	NBR 13528 [11]

suitable application conditions where the coating does not suffer a large volume of water loss.

Granite and marble residues were incorporated in substitution of hydrated lime at 25, 50, 75, and 100% levels beyond the reference without incorporation.

The mortars were made as recommended by NBR 13276. The tests performed and the respective standards are shown in Table 1.

Results

The results of characterization of the materials used in the manufacture of mortar are shown in Fig. 1 where the particle size distribution is illustrated.

The granulometric test shows a great similarity in the distribution of the grain size of marble with that of cement. The granite presented larger grain size than the previous ones, being smaller than the sand.

After the characterization of the materials, the mortars were made. The results of the consistency index tests indicated the amount of water to promote a 260 ± 5 mm spread. The quantities of materials used to make 2.5 kg of mortar are shown in Table 2.

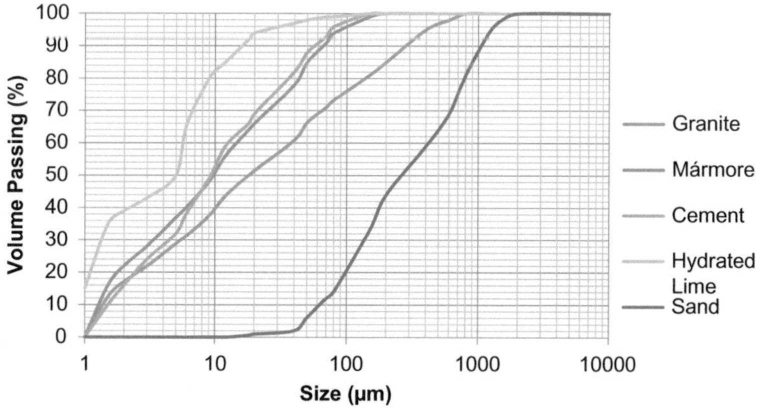

Fig. 1 Granulometry of the used materials

The consistency test results identified a drop in the amount of water as the incorporation content of the residue increased. The drop in the amount of water needed is greater in granite compared to marble. In both traces, the above trends were verified, however, the 1:2:9 trace presents the largest amount of water needed to achieve the same 1:1:6 trace spread. All these trends are justified by granulometry, where hydrated lime is the finest material than cement and waste has greater water absorption, so the traces with the largest amount of hydrated lime are the traces with the most water incorporated into the mixture. Table 2 presents the results of specific mass, unit mass, and moisture absorption.

Figure 2 shows the results of the hardened mortar water absorption test.

The results shown in Fig. 2 show the poor performance obtained by the waste in the water absorption test. Mortars tend to increase water absorption as the amount of residue added increases.

It is noteworthy that up to 50% of marble in both traces, the water absorption showed little significant growth. The results indicate that the grain packing remains similar up to 50% marble, where after this value, the porosity grows significantly.

The granite does not have proper packaging with the materials and thus has increased porosity throughout the increase of incorporation content.

Table 2 Quantity of materials to make 2.5 kg of mortar

Traço		Cement (g)	Hydrated lime (g)	Marble (g)	Granite (g)	Sand (g)	Water (g)
1:1:6	*Reference*	312.50	312.50	0.00	0.00	1875.0	703.0
	25% Marble	312.50	234.38	78.13	0.00	1875.0	696.0
	50% Marble	312.50	156.25	156.25	0.00	1875.0	691.0
	75% Marble	312.50	78.13	234.38	0.00	1875.0	688.0
	100% Marble	312.50	0.00	312.50	0.00	1875.0	680.0
	25% Granite	312.50	234.38	0.00	78.13	1875.0	682.0
	50% Granite	312.50	156.25	0.00	156.25	1875.0	674.0
	75% Granite	312.50	78.13	0.00	234.38	1875.0	661.0
	100% Granite	312.50	0.00	0.00	312.50	1875.0	643.0
1:2:9	*Reference*	208.33	416.67	0.00	0.00	1875.0	828.0
	25% Marble	208.33	312.50	104.17	0.00	1875.0	814.0
	50% Marble	208.33	208.33	208.34	0.00	1875.0	802.0
	75% Marble	208.33	104.17	312.50	0.00	1875.0	796.0
	100% Marble	208.33	0.00	416.67	0.00	1875.0	785.0
	25% Granite	208.33	312.50	0.00	104.17	1875.0	801.0
	50% Granite	208.33	208.33	0.00	208.34	1875.0	784.0
	75% Granite	208.33	104.17	0.00	312.50	1875.0	756.0
	100% Granito	208.33	0.00	0.00	416.67	1875.0	719.0

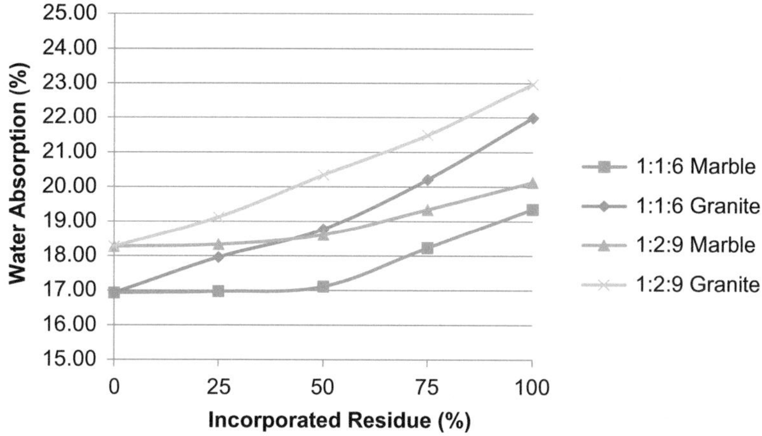

Fig. 2 Water absorption from mortars

Comparing the traces used, the trace with the most hydrated lime and the largest amount of water has the highest porosity, justified by the outflow of water that leaves voids after hardening.

Figures 3 and 4 show, respectively, the results obtained in the flexural tensile strength and compressive strength tests.

Granite, as well as in absorption, shows a decrease in resistance throughout the increase of the incorporation content. Marble has increased strength up to 50% incorporation.

Comparing the results between the traces, it is clear the highest mechanical resistance achieved by the trace with more cement, the 1:1:6 trace.

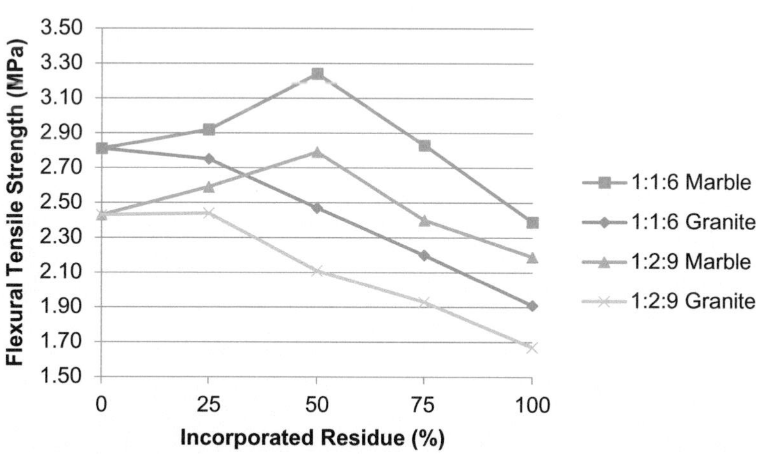

Fig. 3 Flexural tensile strength from mortars

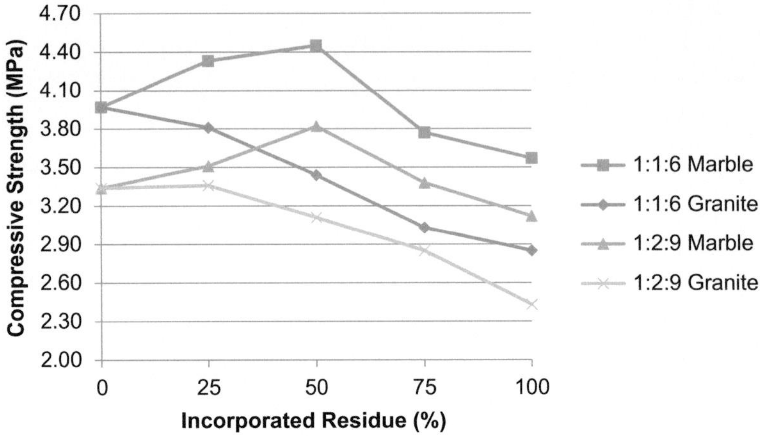

Fig. 4 Compressive strength from mortars

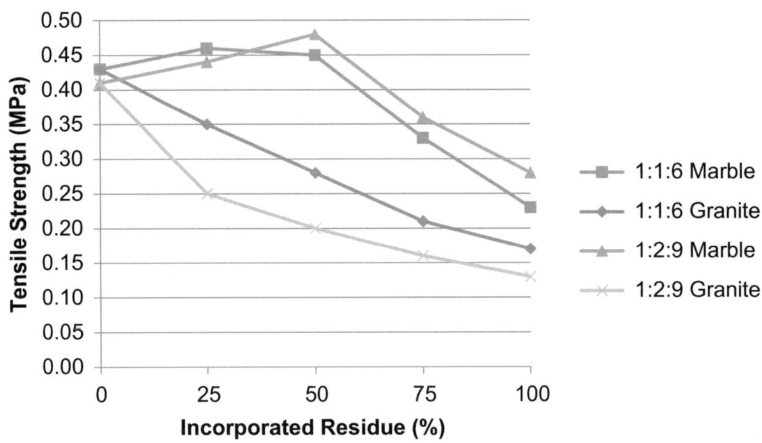

Fig. 5 Tensile strength from mortars

Figure 5 presents the results of the tensile strength test.

It is verified by the test the tendency of increase of the adhesion until the content of 50% of marble incorporation, however, the values present little growth considering the high dispersion of the adhesion results. Increasing marble incorporation above 50% promotes decreased adhesion.

In relation to granite, throughout the granite incorporation content, the adhesion results have decreased.

The results are justifiable considering that the rupture in all samples was in the mortar itself. The mortars were applied on a rough surface and in none of the samples the interface between the mortar and mortar presented poor adhesion.

Considering that the bond between the layers presented no problem and the predominant type of rupture was in the mortar, the tendency of the adhesion result is close to the mechanical resistance of the mortar.

Conclusion

After the results, it can be concluded that

- The characterization of the sands identified large volume of fines and lower specific mass in Waste sand compared to Natural sand.
- Waste sand incorporation decreases the mechanical strength of concrete. The difference is minimum up to 7 days and increases considerably at 28 days. Even after 28 days, waste sand concrete continues to show strong growth and a tendency to approach the final strength of concrete with natural sand.
- Waste sand concretes showed higher water absorption, both by immersion and capillarity. The void content of concrete increases as the incorporation of waste sand increases.

Acknowledgements The authors thank the Brazilian agencies CNPq, CAPES and FAPERJ for the support provided to this investigation.

References

1. dos Santos FF et al (2012) Adequacy of the municipalities of Rio Grande do Sul state to the construction waste management legislation. Iberoam J Ind Eng 4(8):1–18 (In Portuguese)
2. Halmeman MCR, de Souza PC, Casarin AN (2010) Characterization of construction and demolition waste at the solid waste reception unit in Campo Mourão-PR. Revista Tecnológica: 203–209 (In Portuguese)
3. UNEP (2009) Buildings and Climate Change–Summary for Decision-Makers
4. Afonso RG, de Azevedo, Alexandre J, Pessanha LSP, da ST Manhães R, de Brito J, Marvila MT (2019) Characterizing the paper industry sludge for environmentally-safe disposal. Waste Manag 95:43–52
5. Bourscheid AB, Souza RL (2010) Construction and demolition wastes as an alternative material. 1ª edição. Florianópolis, Publicações do IF-SC (In Portuguese)
6. Leite MB (2001) Mechanical properties evaluation of concretes produced with recycled aggregates of construction and demolition waste. Tese de D. Sc, UFRGS, Rio Grande do Sul, Brasil (In Portuguese)
7. Associação Brasileira De Empresas De Limpeza Pública E Resíduos Especiais (Abrelpe). Overview of Solid Waste in Brazil. 2016
8. Hasaba S, Kawamura M, Torli K, Takemoto K (1981) Drying shrinkage and durability of concrete made from recycled concrete aggregates. Jpn Concr Inst 3:55–60

9. Pinto TP (1986) Use of construction waste. Study of the use in mortars. Dissertação de M. Sc., Departamento de Arquitetura e Planejamento da Universidade de São Carlos/USP, São Carlos, SP, Brasil (In Portuguese)
10. Coimbra MA, Libardi W, Morelli MR (2004) Waste preparing of zinc battery for use in mortar and concrete materials of portland cement. Cerâm 50(316):300–307
11. Sormunen P, Kärki T (2019) Recycled construction and demolition waste as a possible source of materials for composite manufacturing. J Build Eng 24

Influence of Construction and Demolition Waste Incorporation in Concrete

Antônio Macedo, Euzebio Zanelato, André Manhães, Afonso Azevedo, Markssuel Marvila, Jonas Alexandre, Sergio Monteiro and Lucio Petrucci

Abstract The construction industry is the human activity that consumes the most natural resources. This occurs from before the beginning of the work, with the extraction of minerals for the production of aggregates and cement. Considering the high consumption of materials in the construction and the durability of the buildings, the continuous generation of CDW (Construction and Demolition Waste) presents itself as a problem. The objective of this work is to verify the influence that the incorporation of CDW promotes in the concrete properties. The CDW was incorporated into 50 and 100% replacement of natural sand. Materials and concrete characterization tests were performed by slump test, water absorption by immersion, water absorption by capillarity and mechanical strength. The tests indicated that the presence of CDW improves the performance of the concrete, increasing its resistance and decreasing its water absorption.

Keywords Construction and demolition · Waste · Concrete

A. Macedo · A. Manhães · A. Azevedo · M. Marvila · L. Petrucci
UCAM – Cândido Mendes University, Av. Anita Peçanha, 100 - Parque São Caetano, Campos dos Goytacazes, Rio De Janeiro 28030-335, Brazil

E. Zanelato (✉) · A. Manhães · M. Marvila · J. Alexandre
LECIV – Civil Engineering Laboratory, UENF - State University of the Northern Rio de Janeiro, Av. Alberto Lamego, 2000, Campos Dos Goytacazes, Rio de Janeiro 28013-602, Brazil
e-mail: ebzanelato@gmail.com

E. Zanelato
DIRINF – Directorate of Infrastructure Rectory, IFF - Federal Institute Fluminese, Rua Cel. Valter Kramer, 357 - Parque Vera Cruz, Campos dos Goytacazes, Rio De Janeiro 28080-565, Brazil

A. Azevedo
TER – Department of Agricultural Engineering and Environment, UFF - Federal Fluminense University, Rua Passo da Pátria, 341, Niterói, Rio de Janeiro 24210240, Brazil

S. Monteiro
Department of Materials Science, IME - Military Institute of Engineering, Square General Tibúrcio, 80, Rio de Janeiro 22290-270, Brazil

B. Li et al. (eds.), *Advances in Powder and Ceramic Materials Science*, The Minerals, Metals & Materials Series,
https://doi.org/10.1007/978-3-030-36552-3_11

Introduction

The Construction Industry deserves to be highlighted in the study of sustainability as one of the most important activities for economic and social development, but it is also a major generator of environmental impacts [1]. Construction activities consume a considerable amount of inert materials such as sand and stones. With the rapid growth of urban regions, the exploitation of natural resources and the generation of construction and demolition waste has reached worrying levels as a result of mismanagement in the works, which generates waste and pollution [2].

Four points are important with regard to sustainability in the sector: the supply chain is long and disjointed, causing ineffective gaps in introducing new technologies; culturally sustainable actions are viewed as costly by builders; disagreements regarding the economic benefits of contractors; and lack of theoretical and practical knowledge and skills on sustainable measures by the population and even by professionals [3].

CDW generation precedes any work. The manufacturing processes of materials used for construction consume natural resources, consume energy, and generate their own waste [4].

According to Bourscheid [5], in construction, there are many losses due to the use of obsolete techniques. SINDUNCON-CE reiterates that waste in a job is the main generator of CDW. These can be caused by overproduction, ineffective stock control, defective manufactured goods, transportation, and processing. The process of unbridled urbanization causes existing buildings to undergo renovation and the increase in population income and housing credit policies has catapulted the emergence of new works and renovations, which further increased waste generation [6].

The CDW was always treated as garbage, expendable, paid for removal from the worksite, without worrying about its destination. 45.1 million tons of CDW were collected in a legal and registered manner in 2015 [7] in addition to a large amount of waste disposed of illegally. Despite its economic potential, concern about the CDW and its reuse and recycling is relatively new, dating back to the 1980s [8, 9]. Figure 1 presents the data of destination of CDW by region of Brazil.

CDW recycling has been successful in several countries [10]. According to Sormunen [11], countries like the USA, Japan, the Netherlands, France, Denmark , and England already have a consolidated industry with hundreds of plants installed.

The CDW recycling chain must be properly managed as it depends on many variables such as technology used, type of waste used, and purpose of recycled material [12].

Many studies have shown the possibility of using recycled aggregates in both the large portion [13, 14] and the small portion [15] for the manufacture of concrete. Leite [6], Corinaldesi [16], and other researchers showed satisfactory performances in fractional substitution of natural aggregates for those recycled in mortars.

The objective of this work is to evaluate the influence that the incorporation of CDW promotes in the performance of concretes.

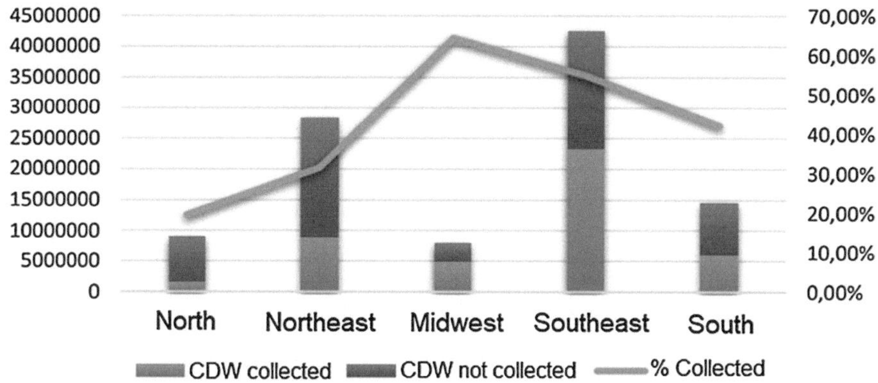

Fig. 1 CDW Collected and CDW not collected by region in Brazil in 2015 Source: ABRELPE, 2016

Materials and Methods

The CDW were manually ground (Fig. 2a) followed by equipment crushing (Fig. 1b) to be stored in suitable containers until use in the assays.

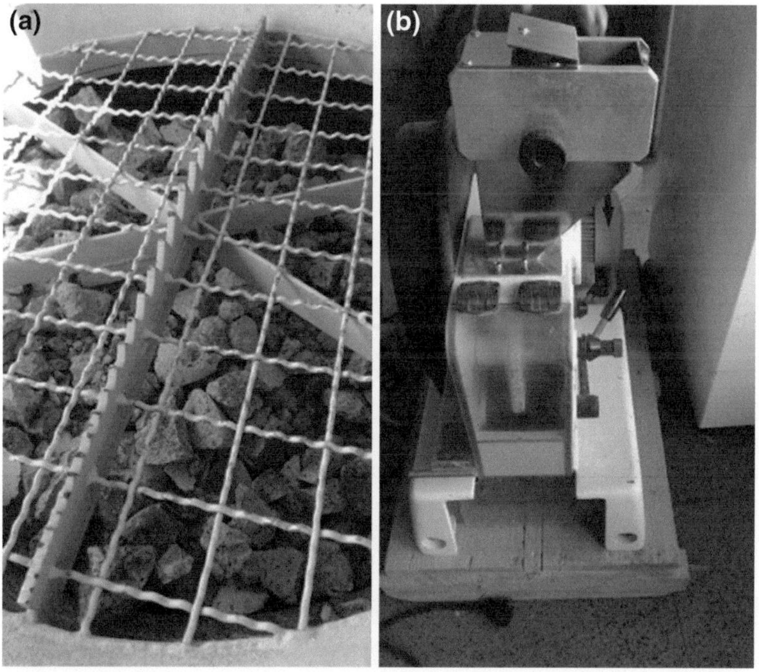

Fig. 2 CDW shredding: **a** Manual, **b** by jaw crusher

Table 1 Concrete traces

	w/c	Cement	Natural sand	Waste sand	Gravel
Ref. (100% NS)	0.46	1	1.32		2.65
100% WS	0.46	1		1.47	2.38
50% NA | 50% WS	0.46	1	0.66	0.74	2.51

The particle size of recycled CDW (WS) and natural sand (NS) sand was verified by sieving tests, according to NBR NM 26 [17] and NBR NM 27 [18].

Specific gravity, grain swelling, bulk density, and moisture absorption tests were also performed for characterization of materials.

The concrete was dosed according to the ABCP method to obtain the resistance of 40 MPa in 28 Days.

Three traces of concrete were prepared using 100% natural sand, 100% waste sand, and 50% natural sand/50% waste sand according to Table 1.

Concrete workability was evaluated by the slump test.

Compressive strength tests were performed at 7, 28, and 56 days according to NBR 5739. The times of 7 and 56 days will be important to follow the resistance growth due to the hydration of the waste sand.

Water absorption and voids index were evaluated according to NBR 9778 [19].

Capillary water absorption was evaluated according to NBR 9779 [20].

Results

Figure 3 shows the results of the particle size of the used sand.

The particle size analysis indicates a higher volume of fines in the waste sand. The largest volume of fines comes from the waste grinding process in which the dust is generated.

Table 2 presents the results of specific mass, unit mass, and moisture absorption.

Waste Sand has lower specific mass and unit mass due to the materials that compose the waste sand. The lower density combined with the higher water absorption of waste sand increases the amount of water needed to maintain concrete workability.

Figure 4 presents the results obtained in the swelling test.

In addition to having a much smaller maximum swelling, the Waste Sand did not have significant swelling up to 5% humidity, on the contrary and retracted in the first water additions behavior can be justified retraction in the first water additions as being caused by an internal cure, provided by water absorption from the aggregates.

Table 3 presents the results obtained in the slump test to verify the concrete workability.

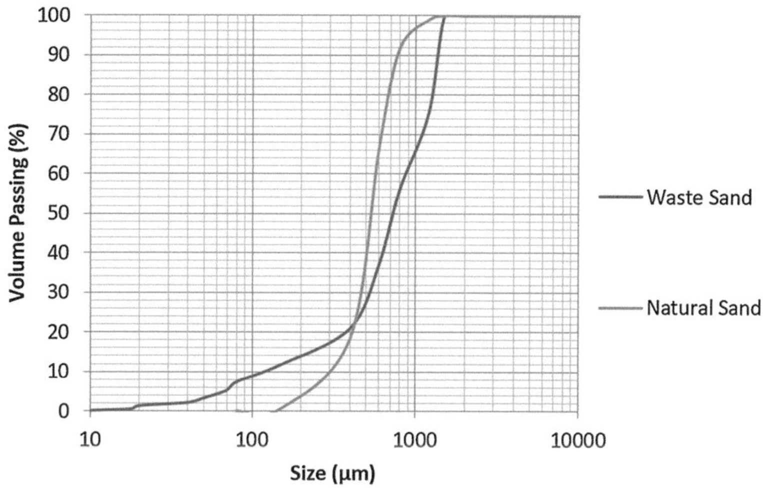

Fig. 3 Granulometry of the used sand

Table 2 Specific mass, bulk density, and moisture absorption results

Material	Specific mass	Bulk density	Moisture absorption
	(kg/m^3)	(kg/m^3)	(%)
Natural sand	2630	1370	0,58
Waste sand	2470	1290	13,72

The decrease in slump as the volume of waste sand increases is justified by the absorption of water from the grains. Part of the mixing water is absorbed by the cement grains present in the waste sand. The lower workability hampers the roll and molding, however, the reduction in workability is not so significant as to preclude their use.

Figure 5 presents the results of the concrete compressive strength test.

The incorporation of waste sand decreases the strength of concrete. The larger the incorporation the greater the reduction.

It is also noteworthy that at 7 days, the resistances of the three traces are similar. At 28 days, the difference between resistances increases considerably and decreases again at 56 days.

This behavior can be explained by the hydration of the waste sand cement that promotes increased resistance at older ages. Despite having lower resistance, the use of Waste sand presents similar results at the beginning and approaching tendency in advanced ages.

Figure 6 presents the results of the Water absorption and Void index.

The results indicate a clear increase in voids index and consequently greater water absorption as the incorporation of CDW was increased. The greater

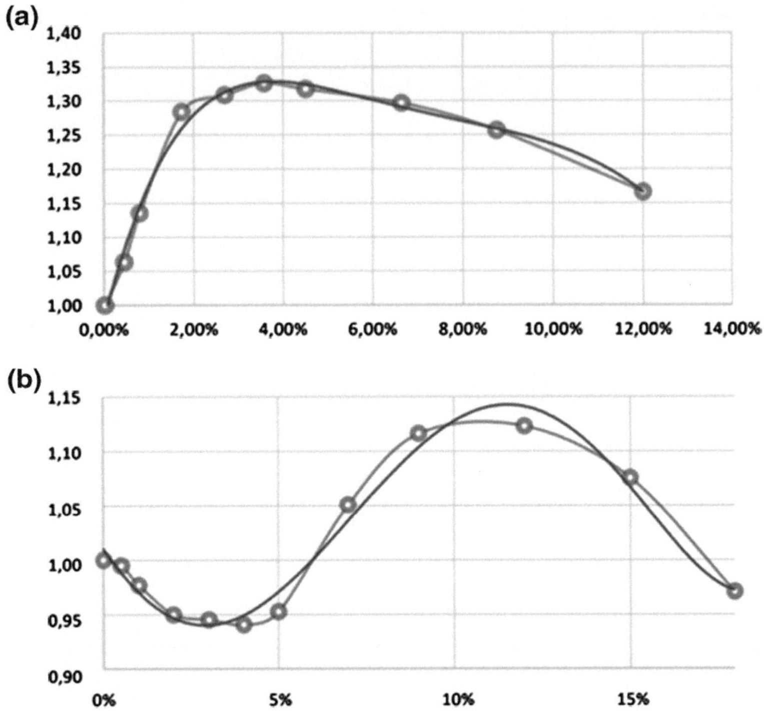

Fig. 4 Swelling test: **a** Natural sand, **b** Waste sand

Table 3 Slump test

Concrete trace	Slump (mm)
100% Natural sand	85
50% Natural sand	71
50% Waste sand	
100% Waste sand	62

absorption of waste sand water by the cement promotes an increase in the voids content and greater absorption capacity of the concrete.

High porosity values, as shown by waste sand concretes, can potentially present durability problems in the structure due to weathering.

Figure 7 presents the results of the Capillarity water absorption test.

Capillary absorption results indicate greater water absorption as the incorporation of the residue is increased, similar to that obtained in the previous assay. This behavior is justified by the type of absorption performed in the capillarity test.

Capillary absorption is highly influenced by active pores with diameters between 0.1 and 10 μm.

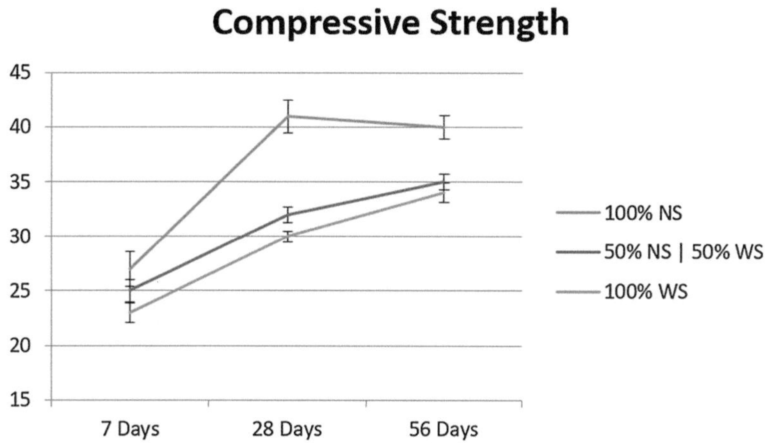

Fig. 5 Compressive strength results

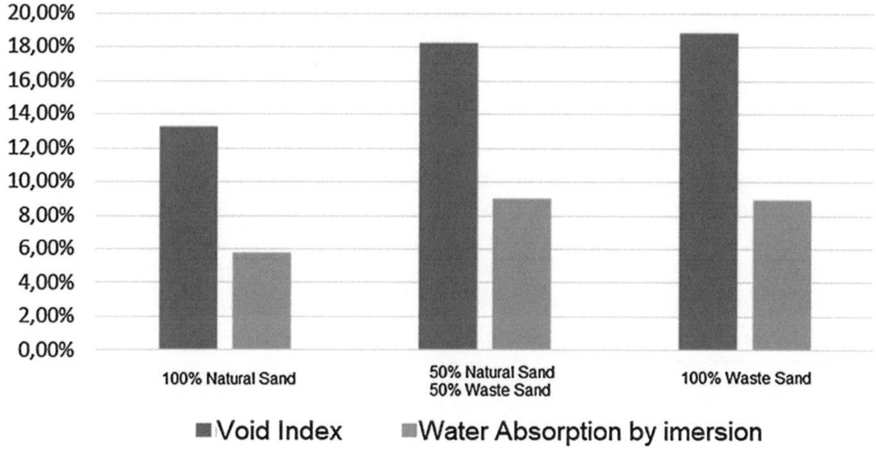

Fig. 6 Water absorption and Void index of concrete

Therefore, besides presenting a large volume of pores of high diameters as verified by the high void index, it also presents a considerable volume of active pores with smaller diameter.

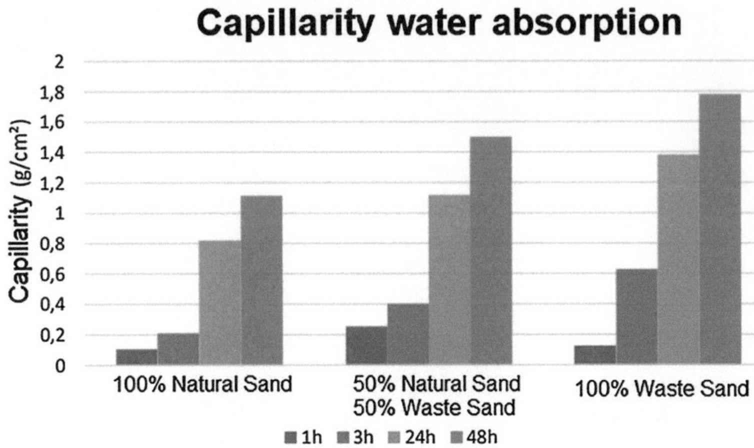

Fig. 7 Capillarity water absorption of concrete

Conclusion

After the results, it can be concluded that

- The characterization of the sands identified a large volume of fines and lower specific mass in Waste sand compared to Natural sand.
- Waste sand incorporation decreases the mechanical strength of concrete. The difference is minimum up to 7 days and increases considerably at 28 days. Even after 28 days, waste sand concrete continues to show strength growth and a tendency to approach the final strength of concrete with natural sand.
- Waste sand concretes showed higher water absorption, both by immersion and capillarity. The void content of concrete increases as the incorporation of waste sand increases.

Acknowledgements The authors thank the Brazilian agencies CNPq, CAPES, and FAPERJ for the support provided to this investigation.

References

1. dos Santos FF et al (2012) Adequacy of the municipalities of Rio Grande do Sul state to the construction waste management legislation. Iberoam J Ind Eng 4(8):1–18 (In Portuguese)
2. Halmeman MCR, de Souza PC, Casarin AN (2010) Characterization of construction and demolition waste at the solid waste reception unit in campo Mourão-PR. Revista Tecnológica 203–209 (In Portuguese)
3. UNEP (2009) Buildings and climate change–summary for decision-makers

4. de Azevedo ARG, Alexandre J, Pessanha LSP, da S.T. Manhães R, de Brito J, Marvila MT (2019) Characterizing the paper industry sludge for environmentally-safe disposal. Waste Manag 95:43–52
5. Bourscheid AB, Souza RL (2010) Construction and demolition wastes as an alternative material. 1ª edição. Florianópolis, Publicações do IF-SC, 2010. (In Portuguese)
6. Leite MB (2001) Mechanical properties evaluation of concretes produced with recycled aggregates of construction and demolition waste. Tese de D. Sc, UFRGS, Rio Grande do Sul, Brasil (In Portuguese)
7. Associação Brasileira de Empresas De Limpeza Pública E Resíduos Especiais (ABRELPE) (2006) Overview of Solid Waste in Brazil
8. Hasaba S, Kawamura M, Torli K, Takemoto K (1981) Drying shrinkage and durability of concrete made from recycled concrete aggregates. Jpn Concr Inst 3:55–60
9. Pinto TP (1986) Use of construction waste. Study of the use in mortars. Dissertação de M. Sc., Departamento de Arquitetura e Planejamento da Universidade de São Carlos/USP, São Carlos, SP, Brasil. (In Portuguese)
10. Coimbra MA, Libardi W, Morelli MR (2004) Waste preparing of zinc battery for use in mortar and concrete materials of portland cement. Cerâmica 50(316):300 307
11. Sormunen P, Kärki T (2019) Recycled construction and demolition waste as a possible source of materials for composite manufacturing. J Build Eng 24
12. Islam R, Nazifa TH, Yuniarto A, Uddin AS, Salmiati S, Shahid S (2019) An empirical study of construction and demolition waste generation and implication of recycling. Waste Manage 95
13. Oikonomou ND (2005) Recycled concrete aggregates. Cement Concr Compos 27:315–318
14. Etxeberria M, Vázquez E, Mari A, Barra M (2007) Influence of amount of recycled coarse aggregates and production process on properties of recycled aggregate concrete. Cement Concr Res 37:735–742
15. Evangelista L, Brito J (2007) Mechanical behaviour of concrete made with fine recycled concrete aggregates. Cement Concr Compos 29:397–401
16. Corinaldesi V, Moriconi G (2009) Behaviour of cementitious mortars containing different kinds of recycled aggregate. Constr Build Mater 23:289–294
17. Associação Brasileira de Normas Técnicas. NBR NM 26: Fine aggregate: Sampling. Rio de Janeiro, 2009
18. Associação Brasileira de Normas Técnicas. NBR NM 27: Aggregate: field sample reduction for laboratory testing. Rio de Janeiro, 2001
19. Associação Brasileira de Normas Técnicas. NBR 9778: Hardened mortar and concrete-Determination of water absorption by immersion - Voids index and specific mass. Rio de Janeiro, 2009
20. Associação Brasileira de Normas Técnicas. NBR 9779: Hardened Mortar and Concrete-Determination of Capillary Water Absorption. Rio de Janeiro, 2012

Preparation of $Na_4V_2O_7$ Powder by Solid-State Reaction

Guishang Pei, Junyi Xiang, Zhongci Liu, Dapeng Zhong, Feifei Pan and Xuewei Lv

Abstract As one of the critical intermediate compounds of the sodium roasting converter slag, sodium pyrovanadate ($Na_4V_2O_7$) powder was synthesized by solid-state reaction using $NaCO_3$ and V_2O_5 as raw materials in this study. The preparation was first evaluated by thermodynamic software FactSage® with the minimum Gibbs free energy principle. Effect of temperature *(T)* and partial pressure of carbon dioxide $P(CO_2)$ was analyzed, and the results indicated that the reaction proceeds extensively with increasing temperature and reducing $P(CO_2)$. TG-DSC was applied to further characterize the preparation process, and it can be found that the reaction proceeds extensively near 540 °C corresponding to carbon dioxide gas escaping. Non-isothermal kinetics with a single scan rate was applied to the solid-state reaction, the average apparent activation energy was obtained using Freeman–Carroll method, equal to 102 ± 6 kJ/mol by mathematic fitting. In addition, XRD further verified the phase composition of $Na_4V_2O_7$, and a large number of voids were detected from SEM images caused by the gas release.

Keywords $Na_4V_2O_7$ · Thermodynamic analysis · TG-DSC · Apparent activation energy

X. Lv (✉)
The State Key Laboratory of Mechanical Transmissions, Chongqing University, No. 174 Shazheng Street, Shapingba District, Chongqing 400044, China
e-mail: lvxuewei@163.com

G. Pei · J. Xiang (✉) · Z. Liu · D. Zhong · F. Pan · X. Lv
College of Materials Science and Engineering, Chongqing University, No. 174 Shazheng Street, Shapingba District, Chongqing 400044, China
e-mail: xiangjunyi126@126.com

Chongqing Key Laboratory of Vanadium-Titanium Metallurgy and New Materials, Chongqing University, Chongqing 400044, China

B. Li et al. (eds.), *Advances in Powder and Ceramic Materials Science*, The Minerals, Metals & Materials Series,
https://doi.org/10.1007/978-3-030-36552-3_12

Introduction

Metal vanadium (V) is frequently used in the metallurgical and chemical industries in the past few years due to its special physical and chemical properties. Almost 88% of the metal vanadium in the world is produced from the vanadium titano-magnetite ore [1–3]. One of the typical processes for extracting vanadium from vanadium slag, which is obtained by oxidizing hot metals that bear vanadium, is sodium roasting followed by water leaching. Sodium carbonate (Na_2CO_3) is used in the extraction of converter slag bearing vanadium. Vanadium slag after roasting process with Na_2CO_3 is regarded as the V_2O_5–Na_2O–CaO–MgO–Fe_2O_3–Al_2O_3–MnO–Cr_2O_3–TiO_2 multiple oxide system, and the V_2O_5–Na_2O binary system is the critical and essential sub-system [4–7]. As one of the intermediate compounds in the V_2O_5–Na_2O binary system, sodium metavanadate ($NaVO_3$), its high-temperature heat capacity and phase transition kinetics have been studied in our previous study [8].

With regard to other intermediate compounds, attention on the property of the sodium pyrovanadate ($Na_4V_2O_7$) should be extensively received to better understand the sodium roasting process. As for $Na_4V_2O_7$, which was reported for the first time by Roscoe [9] in 1870. Then, Bjoernberg et al. [10] proposed that the crystal structure of $Na_4V_2O_7{\cdot}(H_2O)_{18}$ was determined from three-dimensional X-ray diffraction. The crystals are hexagonal, with space group $p6_3/m$, a = 9.2478(5), c = 16.591(2)Å, and the cell contains two formula units. Florenskij et al. [11] presented the patent entitled "Sodium pyrovanadate—is used as colour indicator for high-temperature gas redox medium". Exploring more efficient and economic synthesis methods of $Na_4V_2O_7$ is critical for further fundamental and applied research, only few methods can be found in literature. With regard to solid-state reaction, the indefinite composition of the sodium vanadate products together with the contradictory results given by some authors concerning the structure of vanadium compounds, are calling attention to the mechanism of the reaction between sodium carbonate and vanadium pentoxide. Although Kolta et al. [12] successfully prepared $Na_4V_2O_7$ powder by this method, the apparent activation energy of this solid-station reaction is not well investigated, and it should be further clarified.

Herein, the preparation for $Na_4V_2O_7$ powder was firstly evaluated by thermodynamic calculation. Then, the whole process was continuously recorded by TG-DSC treatment, the frequently-used Friedman–Carroll method for solid-state reaction was exacted in present study to calculate the apparent activation energy based on the mass loss of the whole process. The final products were also analyzed by XRD and SEM.

Materials and Experimental

The powder samples of $NaCO_3$ (purity $\geq$ 99.99%) and V_2O_5 (purity $\geq$ 99.50%) were provided by Aladdin Co., LTD. and LiaoShuo Biological Co., LTD, respectively. The mixtures with specific molar 2:1 for $NaCO_3$ and V_2O_5 were used as the raw materials to synthesize sodium pyrovanadate ($Na_4V_2O_7$) powder through solid-state reaction. The general three-step procedure was introduced into the muffle, the sample was held for 10 h under Argon atmosphere with gas flow of 50 mL/min when the temperature reached at 873 K with heating rate of 10 K/min, the as-prepared sample was then cooled with muffle. The X-ray powder diffraction (XRD) patterns in the 2θ range from 10° to 50° were obtained on a PANalytical X'Pert Powder, Panalytical B.V. (Cu Kα radiation) to investigate the crystalline phases of the products. A scanning electron microscope (SEM, TESCAN VEGA 3 LMII, Czech Republic) was also employed to investigate the morphologies and particle sizes of as-prepared $Na_4V_2O_7$ powder.

Methods

Thermodynamic calculation was used to evaluate the solid-state reaction, which was carried out using Factsage® software with the minimum Gibbs free energy principle. Then, Thermal Analysis Kinetics (TAK) was conducted to further understand the solid-state reaction [13]. The process was also characterized by a TG-DSC (404 F3; Netzsch) at a rate of 10 K/min with protection from high-purity argon at 50 mL/min, the changes of mass and heat flow during the preparation was recorded with detail. The frequently-used Friedman–Carroll method for solid-state reaction was exacted in present study to calculate the apparent activation energy based on the mass loss during the whole process [14–16].

Results and Discussion

Thermodynamic Calculation

The chemical reaction equation for the $Na_4V_2O_7$ formation through solid-state reaction when heating $NaCO_3$–V_2O_5 mixtures in the present study can be described as follows:

$$2Na_2CO_3 + V_2O_5 = Na_4V_2O_7 + 2CO_2 \tag{1}$$

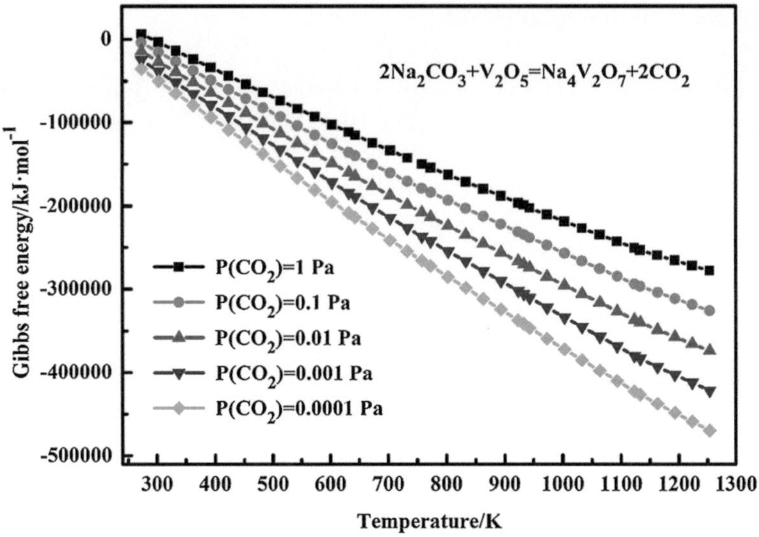

Fig. 1 Gibbs free energy of preparing $Na_4V_2O_7$ powder by solid-state reaction under different *P(CO₂)*

The Gibbs free energy of this reaction with various carbon dioxide pressure (PCO_2) was calculated by the Reaction model of Factsage® software as shown in Fig. 1. It can be found that the Gibbs free energy $P(CO_2)$ is below 0 when carbon dioxide pressure equals 1 Pa, which indicates the reaction can be proceeded when the temperature is beyond 273 K. This solid-state reaction can be speeded byde-creasing the pressure of carbon dioxide.

Non-isothermal Kinetics Analysis

The non-isothermal kinetics analysis was extensively used in the solid-state reaction, and the conversion degree (α) can be defined as follows [17–19]:

$$\alpha_t = \frac{\Delta m_t}{\Delta m0} \quad (2)$$

where Δm_t is the weight loss at specific time t and Δm_o is the theoretical maxium weight loss, according to the solid-state reaction (1), the theoretical maxium weight loss equals 77%.

Figure 2 (left side) shows the changes in the mass and heat flow during the whole process. An obvious endothermic peak accompanied with mass loss of around 6% can be detected when the temperature reacts at near 100 °C, which may be caused by dehydration process of the $NaCO_3$ powder. We carefully operated the

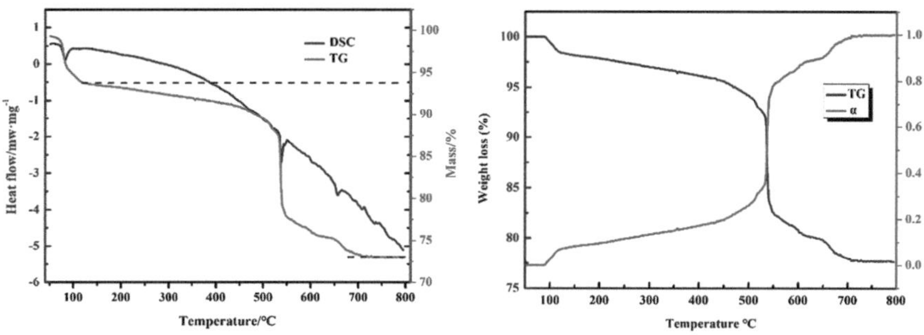

Fig. 2 TG-DSC analysis of preparing $Na_4V_2O_7$ powder by solid-state reaction

experiment, and the addition of water inevitably introduced the mixtures when the raw materials were mixed thoroughly in an agate mortar, and the mass loss of the additional water should be deducted in the subsequent calculation. The mass loss of the mixture gradually increases with increasing the temperature, a large mass loss can be found near 540 °C, indicating that the solid-state reaction proceeds extensively in current stage corresponding to carbon dioxide gas escaping. Afterwards, the mass loss change tends to slow, and the final mass loss approximately equals 21%, which is in great agreement with the calculated value based on chemical reaction equation.

The weight loss and conversion degree are also presented as shown in Fig. 3. The conversion degree increases with the increase in the temperature, the rate of conversion degree reached at the maximum value near 540 °C.

According to the Avrami–Erofeev model [20], the differential equation can be described as follows:

$$\frac{d\alpha}{dt} = k(T)f(\alpha) \tag{3}$$

where $d\alpha/dt$ is the reaction rate (min^{-1}), k(T) is the reaction rate constant (min^{-1}), and $f(\alpha)$ is the model function:

$$\mathrm{K(T)} = \mathrm{Aexp}\left[\frac{-E}{RT}\right] \tag{4}$$

where A denotes the pre-exponential factor, E is the apparent activation energy, and R is the gas constant (8.314 $J \cdot mol^{-1} \cdot K^{-1}$). Put expression (4) into (3), the equation can be obtained as follows:

$$\frac{d\alpha}{dt} = Aexp\left[\frac{-E}{RT}\right]f(\alpha) \tag{5}$$

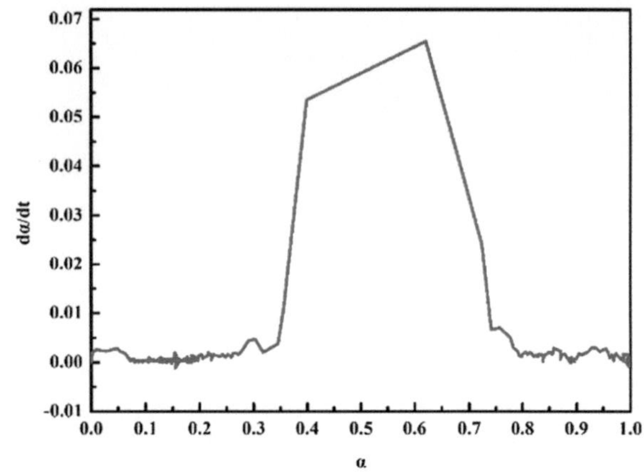

Fig. 3 Weight loss and conservation degree of preparing $Na_4V_2O_7$ powder by solid-state reaction

The apparent activation energy was calculated by Freeman–Carroll method [14–16] in this study and the express also presented as follows:

$$\frac{\Delta ln\left(\frac{d\alpha}{dt}\right)}{\Delta \ln(1-\alpha)} = \frac{-\left(E/R\right)\Delta \frac{1}{T}}{\Delta \ln(1-\alpha)} + n \tag{6}$$

The apparent activation energy can be calculated by plotting the *Δln(1-α)* against *Δ1/T*, the average activation energy equals 102 ± 6 kJ/mol.

Characterization

XRD measurement is the basic analysis method and frequently applied to verify the phase composition evolution during the process of materials preparation and synthesis. The XRD patterns of the final products described and compared with the standard PDF $Na_4V_2O_7$ (NO: 76-1462) are shown in Fig. 4a. By comparing the location and intensity of the diffraction peaks with the standard PDF card, the main phase composition of the presented sample can be checked and verified. The final phase composition was well fitted with the standard PDF cards of $Na_4V_2O_7$. No additional reflection of other phases, e.g., initial reactant, was detected at the final sample. Narrow and strong reflection clearly confirmed good crystallinity of the as-prepared sample.

Surface morphology and microstructure of the as-prepared $Na_4V_2O_7$ powder can be further analyzed by SEM, characteristic micrographs were obtained with four different magnifications and they are given in Fig. 4b. These micrographs reveal the agglomerates of nanoparticles shaped in small chunks of several microns in size. Form the results, we can find that the inner of the prepared sample have few hole which may be caused by the carbon dioxide released during the whole process.

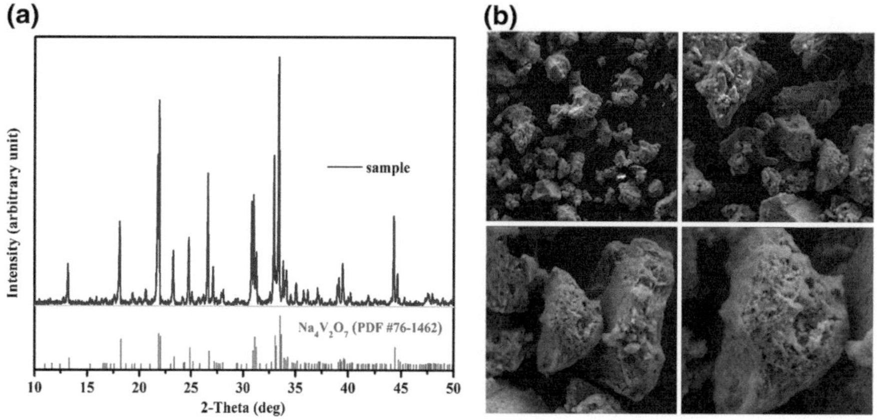

Fig. 4 XRD patterns and SEM images of $Na_4V_2O_7$ powder

A large amount of this hole can extensively accelerate the solid-state reaction by improving the dynamic conditions.

Conclusions

The $Na_4V_2O_7$ powder was prepared by the solid-state reaction. Thermodynamic calculation and non-isothermal kinetics with single heating rate were carried out. The following conclusions were obtained:

1. Thermodynamic calculation indicated that the solid-state reaction of preparing $Na_4V_2O_7$ powder using Na_2CO_3 and V_2O_5 as raw materials, proceed intensively with increasing temperature and reducing partial pressure of carbon dioxide.
2. The solid-state reaction proceeds intensively at 540 °C corresponding to weight loss and endothermic peak, and the final mass loss is in good agreement with the theoretical values based on the chemical reaction equation. The activation energy was obtained by Freeman–Carroll method, the calculated average activation energy equals102 $\pm$ 6 kJ/mol.
3. The XRD further verified the phase composition of preparing $Na_4V_2O_7$ powder, and a large number of voids were detected from SEM images, which probably were caused by the gas release.

Acknowledgements This work was supported by the National Key R&D Program of China (2018YFC1900500), China Postdoctoral Science Foundation (2018M640898), and Graduate Scientific Research and Innovation Foundation of Chongqing, China (Grant No. CYS19001).

References

1. Tavakoli MR, Dreisinger DB (2014) The kinetics of oxidative leaching of vanadium trioxide. Hydrometallurgy 147–148:83–89
2. Xiang JY, Huang QY, Lv XW, Bai CG (2017) Multistage utilization process for the gradient-recovery of V, Fe, and Ti from vanadium-bearing converter slag, J Hazard Mater 3361
3. Xiang JY et al. (2019) A multi-step process for the cleaner utilization of vanadium-bearing converter slag. Paper presented at the 148st TMS Annual Meeting, San Antonio, Texas, 10–14 Mar 2019
4. Kolta GA, Hewaidy IF (1972) Phase diagrams of binary systems vanadium oxide-sodium carbonate and vanadium oxide-sodium sulfate 25(7):327–330
5. Slobodin BV, Fotiev AA (1965) Phase diagram of the Na_2O-V_2O_5 system. Russ J Appl Chem 38(4):801–806
6. Wilson JR, Kerby RC (1973) Solid-liquid phase equilibria for the ternary systems V_2O_5-Na_2O-Fe_2O_3, V_2O_5-Na_2O-Cr_2O_3, and V_2O_5-Na_2O-MgO, Can J Chem 51:1032–1040
7. Pei GS et al (2019) A literature review of heat capacity measurements method. Paper presented at the 148st TMS Annual Meeting, San Antonio, Texas, 10–14 Mar 2019
8. Pei GS, Xiang JY, Lv XW, Li G, Wu SS, Zhong DP, Lv W (2019) High-temperature heat capacity and phase transformation kinetics of $NaVO_3$. J Alloy Compd 794:465–472
9. Roscoe HR (1870) Philos Trans R Soc 160:317
10. Bjoernberg A (1979) Multicomponent polyanions. 24. The crystal structure of $Na_4V_2O_7 \cdot (H_2O)_{18}$. Acta Chem Scand Ser A 33:539
11. Florenskij K, Pkomarov B, Vvolkov V, Pnikolaeva O, Vkudryashova A, Fbashkirova A, Schajkina E A SU736751A1 pyro:vanadium-is used as colour indicator for high-temp. gas redox medium
12. Kolta GA, Hewaidy IF, Felix NS, Girgis NN (1973) Reactions between sodium carbonate and vanadium pentoxide. Thermochim Acta 6(2):165–177
13. Flynn JH (1992) Thermal analysis kinetics-past, present and future. Thermochim Acta 203:519–526
14. Friedman HL (1967) J Macromolecular Sci Chem 41:57
15. Reich L, Levi W (1968) Macromolacular review. Wiley-Interscience, New York, p 173
16. Ma RP, Felder RM, Ferrell JK (1988) Modelling a pilot-scale fluidized bed coal gasification reactor. Fuel Process Technol 19(3):165–290
17. Henderson DW (1979) Thermal analysis of non-isothermal crystallization kinetics in glass forming liquids. J Non-Cryst Solids 30:301–315
18. Ding CY, Lv XW, Chen Y, Bai CG (2016) Crystallization kinetics of $2CaO \cdot Fe_2O_3$ and $CaO \cdot Fe_2O_3$ in the CaO-Fe_2O_3 system. ISIJ Int 56:1157–1163
19. Balamurugan GP, Maiti SN (2010) Nonisothermal crystallization kinetics of polyamide 6 and ethylene-co-butyl acrylate blends. J Appl Polym Sci 107:2414–2435
20. Mccune RC, Wynblatt P (1983) J Am Ceram Soc 66:111

Preparation of $YBO_3:Ce^{3+}$ Film via Impregnation-Lifting Method and Its Photoluminescence Property

Hongen Nian, Xiang Li, Xiaoling Tan, Xiufeng Ren and Jinbo Zeng

Abstract The luminescent YBO_3 thin films with Ce^{3+} doping were successfully prepared by sol–gel and impregnation-lifting method. The phase structure and optical properties were studied. The XRD analysis confirmed the YBO_3 were produced as hexagonal phases at different dosages of Ce^{3+} (x = 0.05, 0.10, 0.15), and the crystallinity of the films was improved by the addition of Ce^{3+}, the luminescence intensity of Ce^{3+} doped YBO_3 film increases with the increase of Ce doping concentration.

Keywords $YBO_3:ce^{3+}$ film · Sol-gel method · Photoluminescence property

Introduction

Conventional fluorescent lamps were usually composed of mixtures of inert gases such as helium and neon, and contain certain amount of mercury vapor, which might release mercury vapor when it is blasted. This type of lamps is certainly not environmental friendly [1]. Numerous studies have focused on the combination of low-concentration metal ions with large band-gap materials to meet the application of vacuum ultraviolet (VUV) excited fluorescent materials in plasma display panels and mercury-free fluorescent lamps through helium discharge [2]. In recent years, rare earth doped nanocrystalline fluorescent materials have received more and more attention. The excellent matrix has an important influence on luminescence properties with the presence of rare earth ions, while the borate matrix shares high

H. Nian (✉) · X. Li · X. Ren · J. Zeng
Key Laboratory of Comprehensive and Highly Efficient Utilization of Salt Lake Resources, Qinghai Institute of Salt Lakes, Chinese Academy of Sciences, Beijing, China
e-mail: nhe@isl.ac.cn

H. Nian · X. Li · X. Ren · J. Zeng
Key Laboratory of Salt Lake Resources Chemistry of Qinghai Province, Xining, China

X. Tan
Qinghai University, Xining, Qinghai, China

B. Li et al. (eds.), *Advances in Powder and Ceramic Materials Science*,
The Minerals, Metals & Materials Series,
https://doi.org/10.1007/978-3-030-36552-3_13

transparency and low melting point [3]. The high chemical stability, low anomalous photo-damage threshold and high solubility of rare earth ions have become an excellent luminescent matrix material [4]. In the borate matrix series, YBO_3 is one of the most excellent luminescent matrix materials. Ce^{3+} doped YBO_3 fluorescent materials absorb the *4f–5d* transition of Ce^{3+} in the near-ultraviolet region to form the blue luminescent peak of the *4f–5d* energy level transition of Ce^{3+}. Compared with $YBO_3:Eu^{3+}$ and $YBO_3:Tb^{3+}$ fluorescent materials, there are few studies on it so far, only high-temperature solid state reactions of $YBO_3:Ce^{3+}$ phosphor have been reported [5–7]. Syu et al. [8] reported the synthesis of $YBO_3:Ce^{3+}$ phosphor by microemulsion precipitation and high-temperature treatment.

In this paper, $YBO_3:Ce^{3+}$ films were successfully synthesized by sol-gel method. The structure and optical properties of $YBO_3:Ce^{3+}$ fluorescent films were studied under different Ce doping concentrations, and the influence of Ce doping on the structure and properties of $YBO_3:Ce^{3+}$ fluorescent films was studied.

Experimental

Chemicals and Materials

The Ce^{3+}-doped yttrium borate film (YBO_3: Ce^{3+}, Ce = 5 wt%, 10 wt%, 15 wt%) was prepared by a sol-gel method on an immersion-drawing coater. $Y(NO_3)_3 \cdot 6H_2O$, $Ce(NO_3)_3 \cdot 6H_2O$, H_3BO_3 were added as starting materials which were dissolved in the water-ethanol solution (1:4), and citric acid and HO $(CH_2CH_2O)_nH$ (PEG) were used as a chelate and cross-linker in the hydrolysis. The resultant solution was stirred on a magnetic stirrer for 5.0 h until it became a clear solution and was aged for 24 h. A piece of silicon sheet was used as the substrate, which was cleaned with deionized water, absolute ethanol, acetone prior to the immersion-drawing process followed by drying and cleaning with an N_2 air gun. The immersion-drawing rate was 1.22 mm/s, and it was dried at 400 °C in a muffle furnace after each coating. Finally, the YBO_3: Ce^{3+} film was annealed at 900 °C.

Characterizations

The phase structure analysis of the film and powders were carried out by XRD (X-ray diffractometer (Model X' TRA, Thermo ARL Corp., USA) to study the phases. The Fourier infrared spectrum was measured on a Fourier infrared spectrometer (Thermo Electron-Nico FIIR) to test functional groups present in the prepared films. The morphology was analyzed by scanning electron microscopy (SEM, HitachTM3000) and atomic force microscope (AFM, Model XE-100E,

PSIA Corp., Korea). The photoluminescence (PL) spectrum was analyzed by fluorescence spectrometry (Model F-4600, Hitachi Corp., Japan) for characterizations.

Results and Discussion

XRD Analysis

The structure of $Y_{1-x}Ce_xBO_3$ (x = 0.05, 0.10, 0.15) films were analyzed by XRD. As shown in Fig. 1, the XRD patterns of $Y_{1-x}Ce_xBO_3$ (x = 0.05, 0.10, 0.15) at different Ce^{3+} doping concentrations are consistent with the XRD patterns at each Ce^{3+} doping concentration, which is characteristic of the hexagonal phase YBO_3. Peaks were all consistent with the standard JCPDS No. 16-0277. No peaks appeared in the XRD patterns of Ce^{3+} at each doping concentration, indicating that the obtained samples were pure phases, and no new phase and other hetero-phases were produced. Meanwhile, the intensity of diffraction peaks increases with the increasing of doping concentration of Ce^{3+}, it indicates that the crystallinity of the films was improved by the addition of Ce^{3+}.

FT-IR Analysis

Figure 2 shows the FT-IR spectrum of $Y_{1-x}Ce_xBO_3$ (x = 0.10) film annealed at 900 °C for 1 h. The test environment is room temperature. The maximum spectral value is 2925 cm^{-1}, which is attributed to the elastic vibration of O-H and C-H bonds from ethanol solution and citric acid. The infrared absorption peak

Fig. 1 XRD patterns of $Y_{1-x}Ce_xBO_3$ (x = 0.05, 0.10, 0.15) thin films annealed at 900 °C for 1 h

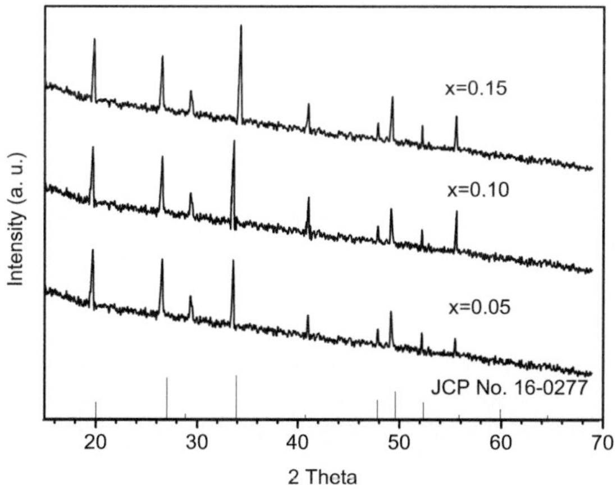

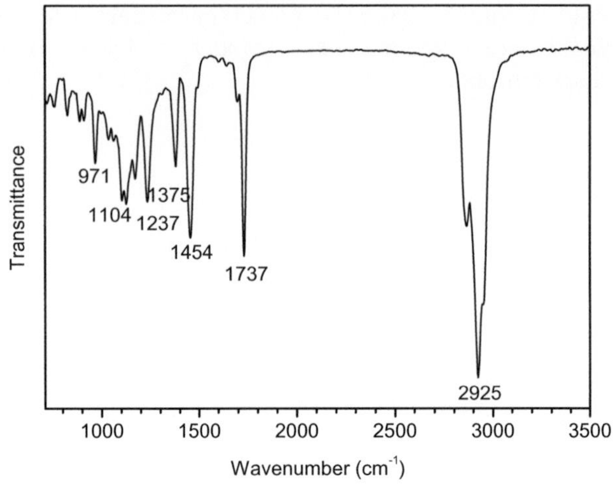

Fig. 2 FT-IR spectrum of the $Y_{1-x}Ce_xBO_3$ (x = 0.10) thin films

of ~1400 cm^{-1} belonged to the typical vibration of the orthogonal phase YBO_3 [9]. In the 1000–1700 cm^{-1} band the IR peaks were indexed to the absorption of carbonate groups (1737 cm^{-1}) and NO_3 group (1454 cm^{-1}), and the bending vibration of hydroxyl groups in water molecules (971 cm^{-1}). The energy band due to the symmetry stretching vibration of the BO bond was also observed. For example, the absorption peak at 1237 cm^{-1} is the anti-symmetric stretching vibration of the BO bond in BO_3^{3-}. The above results indicate that the organic matter and the nitrate are not completely decomposed during the annealing step, and the sample was sustainable under air conditions. Although the organic impurities in the $Y_{1-x}Ce_xBO_3$ (x = 0.10) film samples were still detected in the range of 1000–1200 cm^{-1}, they were found to have no influences on the [BO_4] vibration [10].

Photoluminescence Properties

Figure 3 shows the UV excitation spectra of a $Y_{1-x}Ce_xBO_3$ film monitored at 365 nm. The excitation spectrum of $Y_{1-x}Ce_xBO_3$ film has a broad absorption peak at 356 nm. In the range of 300–380 nm the strongest peak appears at 356 nm, which can be used as the emission spectrum of $Y_{1-x}Ce_xBO_3$ film. The excitation peak can be fitted to two emission peaks at 339 nm and 359 nm by means of the Gaussian deconvolution method, which contributed to the electronic transition inside the BO_3^{3+} groups. The *5d* orbital transition of electrons from O2*p* orbital to B was also observed. The characteristic *4f* emission of Ce^{3+} was also observed in Fig. 3, and the characteristic absorption peak of BO_3^{3+} shows that there is energy transfer between BO_3^{3+} and Ce^{3+}, which was testified by the excitation of 300–339 nm. Absorption bands at 339–359 nm gives indication of the BO_3^{3+}

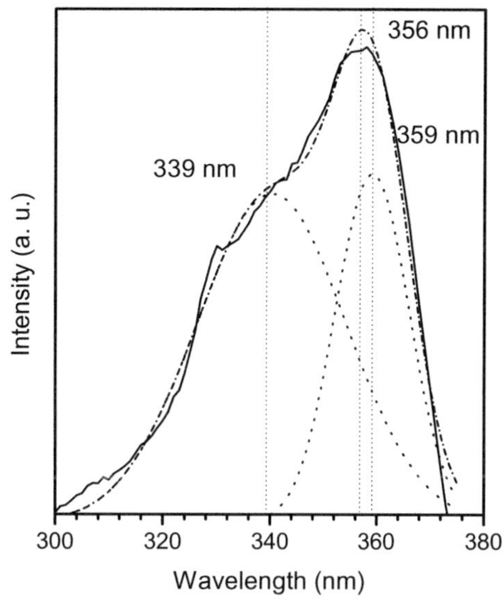

Fig. 3 Excitation spectra of $Y_{1-x}Ce_xBO_3$ thin films

group transfers, its absorbed energy to Ce^{3+}, and this was the typical characteristic absorption by Ce^{3+}.

The emission spectra of $Y_{1-x}Ce_xBO^3$ films treated with different Ce^{3+} doping concentrations after annealing at 900 °C for 1 h at 356 nm show the characteristic *d-f* emission of Ce^{3+} as shown in Fig. 4. The two distinct emission peaks in the figure are located at 369 and 400 nm, belonging to the $^5D_4 \rightarrow {}^7F_6$ transition and the $^5D_4 \rightarrow {}^7F_5$ transition. The maximum luminescence intensity at the $^5D_4 \rightarrow {}^7F_5$ transition can be fitted by Gaussian deconvolution. And two emission peaks were found at 365 and 403 nm, corresponding to the excitation of Ce^{3+} in the lattice of YBO_3. It can be clearly seen from the figure that the doping concentration of Ce

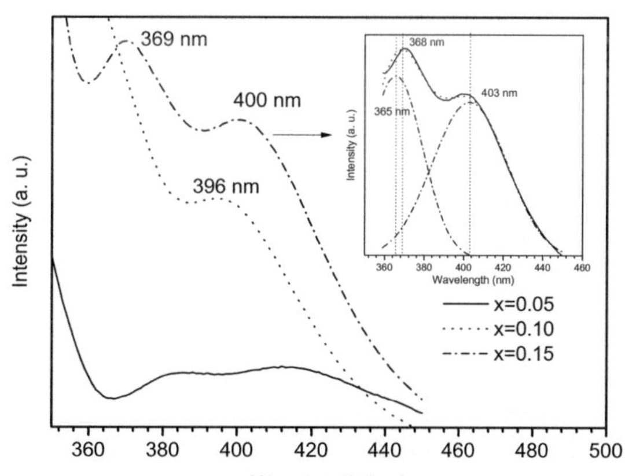

Fig. 4 Emission spectra of $Y_{1-x}Ce_xBO_3$ (x = 0.05, 0.10, 0.15) thin films

significantly affects the luminescence of $Y_{1-x}Ce_xBO_3$ in the form of nanocrystals. The luminescence intensity of $Y_{1-x}Ce_xBO_3$ film increases with the increase of Ce doping concentration. When the doping concentration of Ce is from 5 to 10%, the luminescence intensity of $Y_{1-x}Ce_xBO_3$ film is significantly enhanced to the strongest value. As the Ce doping concentration continues to increase from 10 to 15%, more Ce^{3+} ions entering the $Y_{1-x}Ce_xBO_3$ system lattice are further increased, the Ce^{3+} ion density in the $Y_{1-x}Ce_xBO_3$ system lattice is increased, and this doping level reduces the distance between the Ce^{3+} ions, thereby enhancing the interaction between the Ce^{3+} ions. Under the enhanced interaction between the Ce^{3+} ions, the energy transfer between the Ce^{3+} ions are extremely prone to occur.

Conclusion

The structure of $Y_{1-x}Ce_xBO_3$ ($x = 0.05, 0.10, 0.15$) films is characteristic of the hexagonal phase YBO_3; the obtained samples were pure phases; the intensity of diffraction peaks increase with the increasing of doping concentration of Ce^{3+}. The excitation spectrum of $Y_{1-x}Ce_xBO_3$ film has a broad absorption peak at 356 nm, and two distinct emission peaks are located at 369 and 400 nm. The luminescence intensity of $Y_{1-x}Ce_xBO_3$ film increases with the increase of Ce doping concentration. When the doping concentration of Ce is from 5 to 10%, the luminescence intensity of $Y_{1-x}Ce_xBO_3$ film is significantly enhanced to the strongest value.

Acknowledgements We acknowledge financial support by Qinghai Basic Program under Grant (2017-ZJ-705).

References

1. Feldmann C, Jüstel T, Ronda CR, Schmidt PJ (2003) Inorganic luminescent materials: 100 years of research and application. Adv Func Mater 13:511–516
2. Ronda CR (1995) Phosphors for lamps and displays: an applicational view. J Alloy Compd 225:534–538
3. Park K, Kim J, Kim KY (2012) Enhancement of green emission for Al^{3+}-doped YBO_3:Tb^{3+}. Mater Chem Phys 136:264–267
4. Belsky AN, Krupa JC (1999) Luminescence excitation mechanisms of rare earth doped phosphors in the VUV range. Displays 19:185–196
5. Blasse G, Bril A (1967) Study of energy transfer from Sb^{3+}, Bi^{3+}, Ce^{3+} to Sm^{3+}, Eu^{3+}, Tb^{3+}, Dy^{3+} [J]. J Chem Phys 47(6):1920–1926
6. Blasse G, Bril A (1967) Investigation of some Ce^{3+}-activated phosphors [J]. J Chem Phys 47 (12):5139–5145
7. Zhao X, Wang X, Chen B, Di W, Meng Q, Yang Y (2002) Luminescent properties of YBO_3: RE^{3+}(RE = Ce, Tb) in the UV/VUV region [J]. Proc SPIE Int Soc Opt Eng 6030:6300N–6312N
8. Syu JR, Kumar S, Das S, Lu CH (2012) Microemulsion-mediated synthesis and characterization of YBO_3: Ce^{3+} phosphors [J]. J Am Ceram Soc 95(6):1814–1817

9. Yang L, Zhou L, Chen X, Liu X, Hua P, Shi Y, Yue X, Tang Z, Huang Y (2011) Hydrothermal synthesis of $YBO_3:Tb^{3+}$ microflowers and their luminescence properties. J Alloy Compd 509:3866–3871
10. Onani MO, Okil JO, Dejene FB (2014) Solution–combustion synthesis and photoluminescence properties of $YBO_3:Tb^{3+}$ phosphor powders. Physica B 439:133–136

Sintering Ability of Y-Doped $BaZrO_3$ Refractory with Nano-$CaCO_3$ and the Interaction with Ti_2Ni Alloys

Baobao Lan, Wang Shihua, Yubin Xiao, Xionggang Lu, Guangyao Chen and Chonghe Li

Abstract In this work, the effect of nano-$CaCO_3$ additive on the sinter-ability of Y-doped $BaZrO_3$ refractory and its interaction with Ti_2Ni alloys was studied. The results showed that no second phase was observed in the Y-doped $BaZrO_3$ refractory with nano-$CaCO_3$ additive after sintering at 1750 °C for 6 h. The nano-$CaZrO_3$ promoted the densification and the growth of grains of Y-doped $BaZrO_3$ refractory. The relative density of Y-doped $BaZrO_3$ refractory with nano-$CaCO_3$ addition was about 97.5%. The melting experiment of Ti_2Ni alloys was performed in the Y-doped $BaZrO_3$ crucible with nano-$CaCO_3$ additive at 1650 °C for 5 min, 10 min, and 15 min, respectively. Interaction analysis indicated that the thickness of the erosion layer was 2635 um, 3090 um, and 3689 um, respectively; and the content of oxygen was 0.412 wt%, 0.584 wt%, and 1.140 wt%, respectively.

Keywords Y-doped $BaZrO_3$ · Ti_2Ni · Interface reaction

Intruduction

Titanium alloys have been paid more attention due to their low density, high specific strength and corrosion resistance and for biomedical applications because of their good biological performance [1–4]. The high production cost of titanium

---

B. Lan · Y. Xiao · X. Lu · G. Chen (✉) · C. Li
State Key Laboratory of Advanced Special Steel, Shanghai Key Laboratory of Advanced Ferro Metallurgy, School of Materials Science and Engineering, Shanghai University, Shanghai, China
e-mail: cgybless1@shu.edu.cn

X. Lu · C. Li (✉)
Shanghai Special Casting Engineering Technology Research Center, Shanghai, China
e-mail: chli@staff.shu.edu.cn

W. Shihua
Shanghai University, Shanghai, China

B. Li et al. (eds.), *Advances in Powder and Ceramic Materials Science*, The Minerals, Metals & Materials Series,
https://doi.org/10.1007/978-3-030-36552-3_14

alloys has limited their application. Vacuum induction melting in a refractory crucible is a lower cost production of titanium alloys compared to the conventional induction skull melting in a water-cooled copper crucible. However, the alloy melts with high chemical activity and can easily react with the refractory to contaminate the melts indicating that the refractory should possess the high thermodynamic stability to resist the erosion by the melt. Molten titanium alloys will react with common refractories such as SiO_2, MgO, BN, Al_2O_3. Kartavykh [5] melted Ti–Al-Nb alloy in the BN crucible. A significant interaction occurred between the BN crucible and the molten alloy, and a layer of reactants was attached to the side of the alloy. Frenzel et al. [6] found that the graphite crucible could easily react with the NiTi alloy melt at high temperature as well as the generation of TiC reactant. Economos and Kingery [7] used the ZrO_2 crucible to melt pure titanium alloys. The generation of TiO appeared after melting. Weber [8] used MgO crucible to melt the pure titanium alloy. An obvious interaction region was observed in the crucible side. Calcium oxide was a stable refractory with a low price. However, CaO crucible is easy to hydrate in the air, which would cause the melt contamination by the increase of oxygen concentration during the melting process. Y_2O_3 had higher stability than that of CaO, and the Y_2O_3 crucible was widely used in titanium casting and directional solidification of TiAl alloy. For melting TiAl alloys, the Y_2O_3 crucible exhibited better corrosion resistance than that of CaO, ZrO_2, and Al_2O_3 crucibles [9]. However, the thermal shock resistance of Y_2O_3 crucible was poor and serious breakage would occur after several times of use, which could not realize repetitive use and cause great waste.

Recently, the perovskite compound ($CaZrO_3$ and $BaZrO_3$) was introduced as the refractories for melting titanium alloys because of their high melting point and chemical stability. Zhang et al. [10] melted the NiTi alloy with the $BaZrO_3$ crucible. It was found that the molten alloy did not react with the crucible refractory and the oxygen content of the alloy was only 450 ppm; no refractory elements were found in the alloy. Zhou [11] used the $BaZrO_3$ crucible to prepare the TiFe-based hydrogen storage alloy, which has the hydrogen storage performance and was not weaker than graphite as well as no contamination of carbon. In addition, Chen et al. [12, 13] found that the high titanium content of alloys would react with the $BaZrO_3$ refractory because of the dissolution of $BaZrO_3$ in the titanium alloy melts, and severe damage of the $BaZrO_3$ crucible would happen. Currently, doping is a simple and potentially effective method to improve the stability of $BaZrO_3$, the oxide, such as In_2O_3 [14] and Y_2O_3 [15], which are the common additives. Due to the high activity of titanium melts, some additives, such as the In_2O_3 and Bi_2O_3, can react with and contaminate the melts, obviously, they should be excluded as the additives for melting titanium alloys. Y_2O_3 exhibits sufficient stability to titanium melts and implies that it may be an appropriate additive. Kang et al. [16] revealed that the stability of $BaZrO_3$ refractory could be improved by doping Y_2O_3, and the Y-doped $BaZrO_3$ crucible exhibited a better performance than that of the $BaZrO_3$ crucible after melting the Ti_2Ni alloys. Meanwhile, Chen [17] presented that the CaO also was a good additive to improve the stability of the $BaZrO_3$ refractory. However, the effects of nano-$CaCO_3$ additive into Y-doped $BaZrO_3$ and the interface reaction between Y-doped $BaZrO_3$ crucible with nano-$CaCO_3$ additive and titanium-rich alloys have not been reported.

In this paper, the BZY was doped with 10 mol% nano-$CaCO_3$ for melting Ti_2Ni alloy. The experimental process was smelting Ti_2Ni alloy at 1650 °C for 5 min, 10 min, and 15 min, respectively. The relationship between the thickness of the crucible reaction layer and the smelting time at 1650 °C was studied.

Experiment

First, the industrial grade $BaCO_3$, ZrO_2, and Y_2O_3 were selected according to the molar ratio n($BaCO_3$): n(ZrO_2): n(Y_2O_3) = 1: 0.97: 0.015 for preparing the $BaZr_{0.97}Y_{0.03}O_3$ (BZY) refractory. Then, the powder was mixed in a ball mill for 8 h in a ratio of mass ratio m (powder): m (ball mill beads): m (alcohol) = 2: 3: 0.4, so that they were thoroughly mixed. And the BZY powder was made by reacting the corresponding oven-dried mixtures at 1400 °C for 12 h. Next, the BZY refractory was doped with 10 mol% nano-$CaCO_3$, and the powder was mixed in a ball mill for 8 h to obtain nano-$CaCO_3$-doped BZY (BZY–Ca); the distribution of the powders is shown in Table 1. The powder mixtures were fabricated into the crucibles in a U-shape steel mandrel of 3.5 cm width and 4.5 cm height by using the cold isostatic pressing with a pressure of 120 MP for 3 min and sintered with dwell times of 3 h and 6 h at 1450 °C and 1750 °C, respectively. The crucible after sintering had a width of 2.9 cm and a height of 3.7 cm. A slow heating rate was controlled at 2 °C/min to avoid the generation of cracks.

The objective of our melt trials was to produce Ti_2Ni ingot. High purity raw Ni sheet (> 9.99%) and sponge titanium (> 9.9%) were used as the raw materials. Before the melting, the crucible was placed in the VIM furnace and then backfilled with MgO ramming mass to prevent damage of the induction coil in case of crucible failure. The raw material ingredients were placed in the crucible and the furnace was evacuated to 10^{-3} Pa and backfilled with high purity argon gas for at least three times, in order to isolate the effect of the oxygen pickup exclusively from the chamber. Then the melting was performed under the high vacuum. Once the melt was visible, backfilled the argon gas quickly and raised gradually the temperature up to 1650 °C, and kept at this temperature for 5 min. Then the melt was poured into the graphite crucible and cooled in the furnace. The experiment of melting Ti_2Ni alloys for 10 and 15 min was followed by the same procedure, respectively.

Table 1 the mole ratio of $BaCO_3$, ZrO_2, Y_2O_3 and nano-$CaCO_3$ for preparing the BZY–Ca

	$BaCO_3$	ZrO_2	Y_2O_3	Nano-$CaCO_3$
BZY–Ca	1	0.97	0.015	0.01

The phase identification was conducted by X-ray diffraction (XRD, D8 advance, Bruker) using nickel filtered CuKa radiation. The sidewall of the crucible after melting was investigated using the scanning electron microscope (JSM-6700F). To evaluate the melt contamination of TiAl alloy by the crucible refractory, the alloy was investigated by inductively coupled plasma atomic emission spectrometry (ICP-AES). During the sample preparation, an about 10 mm thick ingot was first removed from the alloy by wire-electrode cutting, then, 0.1 g samples were obtained by abrasive cutting, and washed in the hydrofluoric acid for ICP-AES. Overall oxygen content in the alloy was measured by LECO TC600 O/N analyzer.

Results and Analysis

The densities of the crucibles with nano-$CaCO_3$ additives were determined by full automatics density analyzer (ACCUPYC II 1340). And the density of BZY–Ca crucible is 5.94 g/cm^3, while the theoretical density (6.09 g/cm^3) of Y-doped $BaZrO_3$ with nano-$CaCO_3$ is calculated by the weighted average method [17], the relative density of BZY–Ca crucible is 97.9%. The XRD patterns of the BZY powder and BZY–Ca powder were shown in Fig. 1. The reflections for BZY powder was well indexed with cubic symmetry (Pm-3m) about ICDD data for $BaZrO_3$ [PDF-06-0399]. No traces of second phase were found indicating the substitution of Y at lattice sites of $BaZrO_3$ matrix. Meanwhile, the degree of peak shift of nano-$CaCO_3$-doped BZY (BZY–Ca) refractory appeared. It was because that a shrinkage of $BaZrO_3$ lattice (d-value) after doping the Ca ion would happen due to the smaller atomic radius of Ca^{2+} (0.1 nm) compared with that of Ba^{2+} (0.135 nm). According to the Bragg's law: $2d\sin\theta = n\lambda$, the decreasing of d-value will lead to the increasing of the θ-value.

Figure 2 presents that the second electron (SE) and backscattered electron (BSE) images for the BZY–Ca pellets. No second phase was observed in the

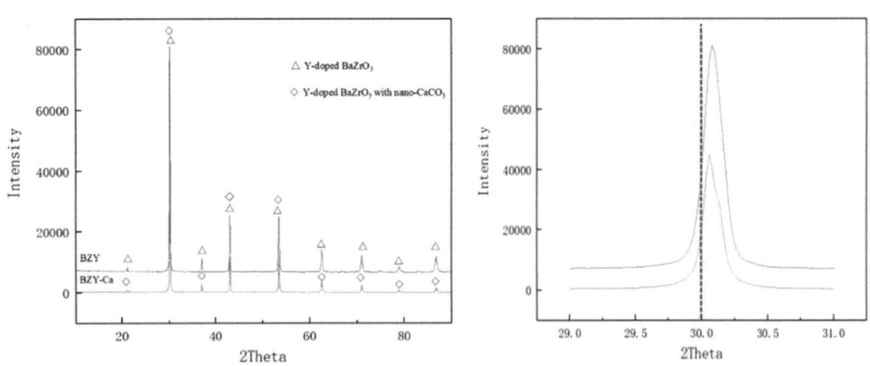

Fig. 1 XRD spectra for the nano-$CaCO_3$-doped BZY refractory

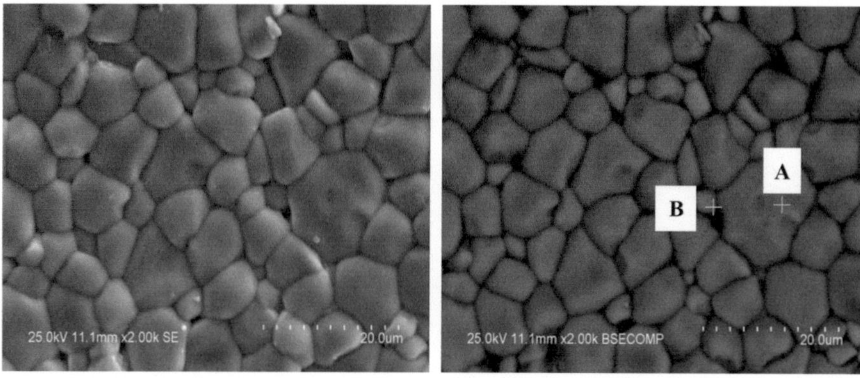

Fig. 2 The second electron (SE) and backscattered electron (BSE) images for the BZY–Ca pellet

BZY–Ca pellet by combining with the XRD spectra in Fig. 1. Apparently, the Ca element could solute into the BZY refractory. Combined with the result of the EDS of the crucible, it can be concluded that the pellet was contained only the BZY–Ca phase (Table 2).

The Ti_2Ni alloy was melted in the BZY–Ca crucibles at 1650 °C for 5 min, 10 min, and 15 min, respectively. The morphology and corrosion of the crucibles after smelting was shown in Fig. 3. Finally, the contamination of the alloy after different smelting times was investigated. As the melting time was increased, the corrosion degree of the crucible by the alloy melt was also gradually increased. As shown in Fig. 3, the thickness of corrosion layer is 2635 um, 3090 um, and 3689 um, when the melting time was 5 min, 10 min, and 15 min, respectively. The contents of Zr, Ba, Ca, O elements in the alloy were shown in Table 3. From the Table 3, the impurities in the Ti_2Ni alloys were also increased with the increasing melting time. The content of oxygen was increasing from 0.412 wt% to 1.14 wt%. Kang [18] presented that the dissolution of Y-doped $BaZrO_3$ refractory in the titanium alloy melt was the main responsibility for the corrosion of the crucible refractory. Apparently, the Y-doped $BaZrO_3$ refractory with nano-$CaCO_3$ additive also exhibited a high thermodynamic stability to the titanium alloy melts. Although the inside of the crucible was eroded, the morphology of the crucible was also retained. Meanwhile, the dissolution content of the refractory was increased with the increasing melting time. According to the previous study in our group, the $BaZrO_3$ refractory could be dissolved in the titanium melts leading to the erosion of

Table 2 The EDS results of point A and B in Fig. 2

Position	Element/at. %					Possible phase
	Ba	Ca	Zr	O	Y	
A	31.38	4.17	29.29	31.58	3.59	$Ba_xCa_{1-x}Zr_yY_{1-y}O_3$
B	32.59	4.14	29.32	30.35	3.60	$Ba_xCa_{1-x}Zr_yY_{1-y}O_3$

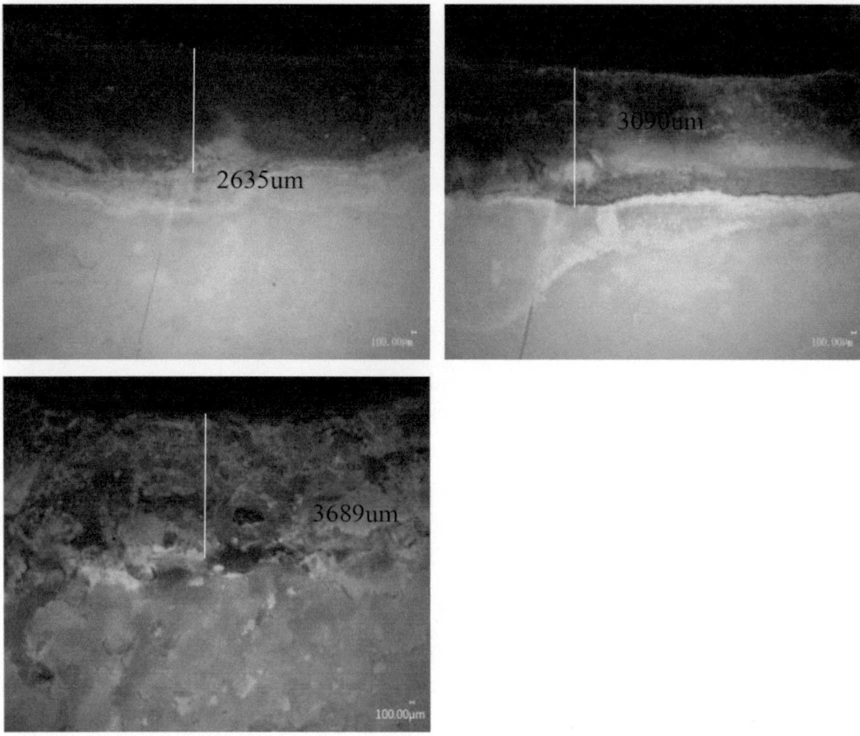

Fig. 3 The macro picture of the corrosion layer of the crucible after melting at 1650 °C for 5 min, 10 min and 15 min respectively **a**: 5 min; **b**: 10 min; **c**: 15 min

Table 3 The concentration of impurities in the Ti_2Ni alloy after melting

	Time	Ba (ppm)	Zr (wt%)	Ca (wt%)	O (wt%)
BZY–Ca	5	18	0.85	0.019	0.414
	10	16	1.14	0.018	0.584
	15	6	1.07	0.019	1.14

the refractory [19], Kang [20] used BZY crucible to melt Ti_2Ni alloy and found that the crucible does not react with Ti_2Ni alloy; it may be caused by decomposition of refractory. Y-doped $BaZrO_3$ refractory and Ca-doped $BaZrO_3$ refractory will dissolve in the titanium melt, and dissolve erosion process is directly related to the Gibbs free energy of formation for the refractory; the Gibbs free energy of formation for the $BaZrO_3$ refractory could decline further with Ca and Y co-additive, so the dissolution of Ca–Y doped $BaZrO_3$ refractory will occur in the process of titanium melting.

Conclusion

(1) No secondary phase was observed in the Y-doped $BaZrO_3$ refractory with nano-$CaCO_3$ additive, and the refractory was a cubic perovskite structure. Besides, the relative density of the crucible was 97.8%.
(2) After melting the Ti_2Ni alloy at 1650 °C for 5 min, 10 min, and 15 min, respectively, the thickness of the erosion layer of the crucible was 2635 um, 3090 um, and 3689 um, respectively.

Acknowledgements The authors thank the National Natural Science Foundation of China-CHINA BAOWU STEEL GROUP Joint research fund for iron and steel (No.: U1860203, U1860108); National Natural Science Foundation of China (No.: U1760109); Scientific research innovation project of Shanghai education commission (No.:15ZS030).

References

1. Wang X D, Lu FS, Jia H, et al(2014) Titanium industry development report of China in 2013. China's Titanium Industry 21(1):30–36 (in Chinese)
2. Wu X (2006) Review of alloy and process development of TiAl alloys. Intermetallic 14 (10):1114–1122
3. Boyerr RR (1996) An overview on the use of titanium in the aerospace industry. Mater Sci Eng A 213(1–2):103–114
4. Niinomi M, Boehlert CJ (2006) Titanium alloys for biomedical applications. Mater Sci Eng C 26(8):1269–1277
5. Kartavykh AV, Tcherdyntsev VV, Zollinger J (2009) TiAl–Nb melt interaction with AlN refractory crucibles. Mater Chem Phys 116(1):300–304
6. Frenzel J, Zhang Z, Neuking K et al (2004) High quality vacuum induction melting of small quantities of NiTi shape memory alloys in graphite crucibles. J Alloy Compd 385(1):214–223
7. Economos G, Kingery WD (2010) Metal-ceramic interactions: ii, metal-oxide interfacial reactions at elevated temperatures. J Am Ceram Soc 36(12):403–409
8. Weber BC, Thompson WM, Bielstein HO et al (2010) Ceramic crucible for melting titanium. J Am Ceram Soc 40(11):363–373
9. Tetsui T, Kobayashi T, Mori T et al (2010) Evaluation of yttria applicability as a crucible for induction melting of TiAl alloy. Mater Trans 51(9):1656–1662
10. Zhang Z, Zhu KL, Liu LJ et al (2013) Preparation of $BaZrO_3$ crucible and its interfacial reaction with molten titanium alloys. J Chin Ceram Soc 41(9):1278–1283 (in Chinese)
11. Li CH, Zhou H, Chen GY et al (2016) Preparation of TiFe based alloys melted by $BaZrO_3$ crucible and its hydrogen storage properties. J Chongqing Univ 39(2):107–113 (in Chinese)
12. Chen GY, Cheng ZW, Wang SS et al (2016) Interfacial reaction between high reactivity titanium melt and BaZrO3 refractory. J Chin Ceram Soc 44(6):890–895
13. Chen GY, Li BT, Zhang H et al (2017) Corrosion mechanism of BaZrO3 refractory in titanium enrichment melt. Chin J Nonferrous Metals 27(5):947–952
14. Bi L, Fabbri E, Sun Z et al (2011) Sinteractivity, proton conductivity and chemical stability of $BaZr_{0.7}In_{0.3}O_{3-\delta}$ for solid oxide fuel cells (SOFCs). Solid State Ion 196(1):59–64
15. Reddy GS, Bauri R (2016) Y and In-doped BaCeO3–BaZrO3 solid solutions: chemically stable and easily sinterable proton conducting oxides 688:1039–1046
16. Kang JY, Chen GY, La BB et al (2018) Preparation of high density $BaZr_{0.97}Y_{0.03}O_{3-\delta}$ ceramic and its interaction with titanium melt. Key Eng Mater 768:261–266

17. Chen GY, Kang JY, Gao PY et al (2018) Effect of CaO additive on the interfacial reaction between the $BaZrO_3$ refractory and titanium enrichment melt. Rare metal technology 2018. In: The minerals, metals & materials series. https://doi.org/10.1007/978-3-319-72350-1_22
18. Kang JY, Chen GY, Lan BB et al (2019) Stability of Y-doped $BaZrO_3$ and resistance to titanium melt. Chin J Nonferrous Metals 29(4):749–754 (in Chinese)
19. Chen GY, Kang JY, Gao PY et al (2018) Dissolution of $BaZrO_3$ refractory in titanium melt[J]. Int J Appl Ceram Technol. https://doi.org/10.1111/ijac.13009
20. Kang JY, Chen GY, Lan BB et al (2019) Sintering behavior of Y-doped BaZrO3 refractory with TiO2 additive and effects of its dissolution on titanium melts[J]. Int J Appl Ceram Technol 16(3):1088–1097

Synthesis of Silicate Zinc Bioceramic via Mechanochemical Technique

Sorour Sadeghzade, Rahmatollah Emadi and Fariborz Tavangarian

Abstract Hardystonite is currently recognized as a biocompatible bio-ceramic material for a range of medical applications. In this study, pure nanocrystalline hardystonite powder was prepared by mechanochemical synthesis of zinc oxide, silicate oxide, and egg shell in a planetary ball mill followed by sintering. It was found that pure nanocrystalline hardystonite powder formation occurred following 20 h of milling and subsequent sintering at 1000 °C for 3 h. Hardystonite scaffold was prepared by space holder method. The results showed that 3D porous scaffolds with pore sizes in the range of 200–300 μm, total and open porosity of 81 and 76%, respectively, with compressive strength and modulus of 0.35 and 10.49 MPa, were obtained. The average crystallite size of the prepared hardystonite powder and scaffold was measured to be 28 ± 3 and 79 ± 1 nm, respectively. The bioactivity of prepared scaffold was evaluated by simulated body fluid (SBF). Considering the results obtained, it seems that, manufactured scaffolds could be a good candidate for bone tissue engineering applications.

Keywords Hardystonite · Bioactivity · Nanostructure · Scaffolds

Introduction

Recently, using ceramics in biomedical application have attracted many interests due to their excellent biological properties such as bioactivity and biocompatibility [1–3]. Despite this, ceramics have some disadvantages, such as brittleness, causing a low tensile strength. Ceramics can be used in bone substitution due to the ideal

---

S. Sadeghzade · R. Emadi
Materials Research Group, Department of Materials Engineering,
Isfahan University of Technology, 84156-83111 Isfahan, Iran

S. Sadeghzade · F. Tavangarian (✉)
Mechanical Engineering Program, School of Science, Engineering and Technology,
Pennsylvania State University, Harrisburg, Middletown, PA 17057, USA
e-mail: f_tavangarian@yahoo.com

B. Li et al. (eds.), *Advances in Powder and Ceramic Materials Science*,
The Minerals, Metals & Materials Series,
https://doi.org/10.1007/978-3-030-36552-3_15

properties such as good bioactivity, biocompatibility, cell attachment, and non-cytotoxity [3–6]. Based on the literature, calcium phosphate ceramics have possible health threats and more risk assessment is required [7, 8]. The silicate bioceramics are superior to calcium phosphate ceramics in terms of biological properties. Silicate ceramics can promote osteogenic differentiation of stem cells by releasing Si- containing ionic products [7–9].

Zinc silicate ceramics, such as willemite and hardystonite are superior to calcium silicate ceramics due to their higher bond strength [10, 11]. According to Zreiqat et al. [11], the zinc ion plays an important role in tissue regeneration, due to its ability to associate with many proteins. It can lead to a lower dissolution rate. Hardystonite also has a more controllable rate of degradation than that of zinc silicates [10–14].

Sadeghzade et al. [13] reported the bio behavior of hardystonite synthesized by two step sintering method (sintered at 1100 °C for 30 min followed by 900 °C for 13 h). They found excellent biocompatibility and mechanical properties matching to that of human bone. The compressive strength (121 ± 2 MPa) and young modulus (27 ± 4 GPa) of the produced hardystonite were similar to that of cortical bone. As a result, hardystonite is an ideal candidate to promote bone density.

The purpose of this study was to develop a simple and economical method to synthesize pure hardystonite nanopowder. Due to the desirable properties of nanosized bioceramic scaffolds such as high surface area to volume, high contact area, improved adhesion and cellular dissociation, increased apatite formation and bonding to tissue, the hardystonite scaffold with nanostructure was fabricated. For this purpose, hardystonite scaffold was prepared by space holder method. The microstructure, size, shape, and amount of porosity, mechanical and bioactivity properties of scaffolds were evaluated to observe its potential for using in tissue engineering applications.

Material and Method

Hardystonite nanopowder was prepared by mechanochemical synthesis. Zinc oxide (ZnO, 99% purity, Merck), egg shell, and silicate oxide (SiO_2, 99% purity, Aldrich) powders, were mixed as initial materials. The mixtures were put in zirconia vial containing five zirconia balls with the ball/powder mass ratio of 10:1. After 20 h ball milling the powder was sintered at 1000 °C for 3 h.

To prepare hardystonite scaffolds with a nanostructure, pure hardystonite powder was mixed with 85 wt% NaCl as spacer. The powder mixtures were uniaxial pressed into pellets in a hardened steel mold at a pressure of 65 MPa by a universal testing machine (HOUNSFIELD: H50KS, crosshead rate of 0.5 mm/min). The green bodies were then sintered at 1250 °C for 3 h [15].

In order to characterize the powder and scaffolds phase transformation, surface morphology of samples before and after modification, X-ray diffractometry (XRD, X' pert Philips) with CuKα radiation (λ = 0.154 nm) and scanning electron microscopy at an acceleration voltage of 30 kV were used. Also, transmission electron microscopy (TEM) was utilized to evaluate the morphology and grain size of the prepared powders. Mechanical properties of the scaffolds were determined by the universal testing machine (Hounsfield, H25KS). The diameter and height of samples were chosen to be 10 × 20 mm. Also, the porosity of samples was measured by Archimedes' principles [15]. Bioactivity of scaffolds was performed according to Bohner and Limatre instruction [16]. The scaffolds with height and diameter of 3 × 10 mm were immersed in SBF with pH of 7.4 and at a temperature of 37 °C. The pH changes of SBF solution were measured every 2 days up to 21 days.

Results and Discussion

Figure 1 shows the XRD pattern of the hardystonite nanopowder after 20 h of ball milling time with subsequent annealing at 1000 °C for 1 h. The absence of zinc compounds as impurities such as willemite at this temperature can be attributed to the effect of ball milling on the homogenization of the raw materials and increasing their surface area. According to Williamson–Hall method the crystallite size of hardystonite powder was measured to be 21 ± 3 nm.

The XRD pattern of pure hardystonite scaffold was illustrated in Fig. 1. As can be seen, no impurity phase was observed as a result of the reaction of sodium chloride with hardystonite powder. While no peak of sodium chloride was detected

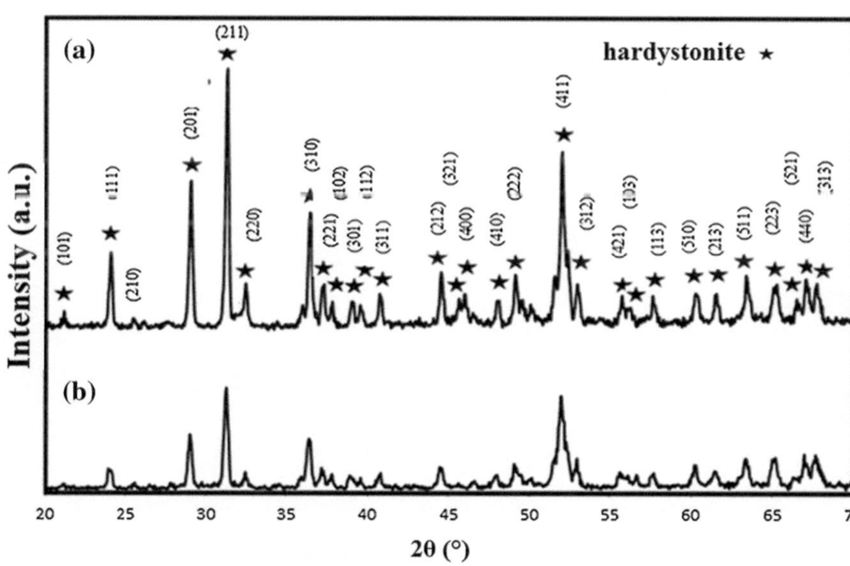

Fig. 1 Hardystonite **a** scaffold and **b** powder

in the structure, which proved the complete removal of spacers during the sintering process.

The increase in the hardystonite peaks can be attributed to the high sintering temperature (1200 °C) and the removal of the spacer particles to produce a porous scaffold. Also, the crystallite size of hardystonite scaffold was measured to be 79 ± 1 nm.

Figure 2 shows the SEM and TEM images of hardystonite nanopowder and scaffold with 81% total porosity at different magnification. Based on Fig. 2a and b, the mean crystallite and agglomerated particle size were measured to be 28 ± 3 and 191 ± 3 nm, respectively.

Figure 2c and d show the morphology, size, and interconnected porosities in hardystonite scaffold. As can be seen in Fig. 2c and d, interconnected pores with the size in the range of 200–300 μ are visible. Such a structure is important in biological scaffolds. In fact, the pores with micron size can allow the tissue and vessel growth inside of the pores as well as allowing the nutrients and proteins to be delivered to the cells leading to the replacement of the scaffold with new tissues. Therefore, the above scaffold will be appropriate for tissue engineering applications. Table 1 shows the mechanical and physical properties of the hardystonite scaffold.

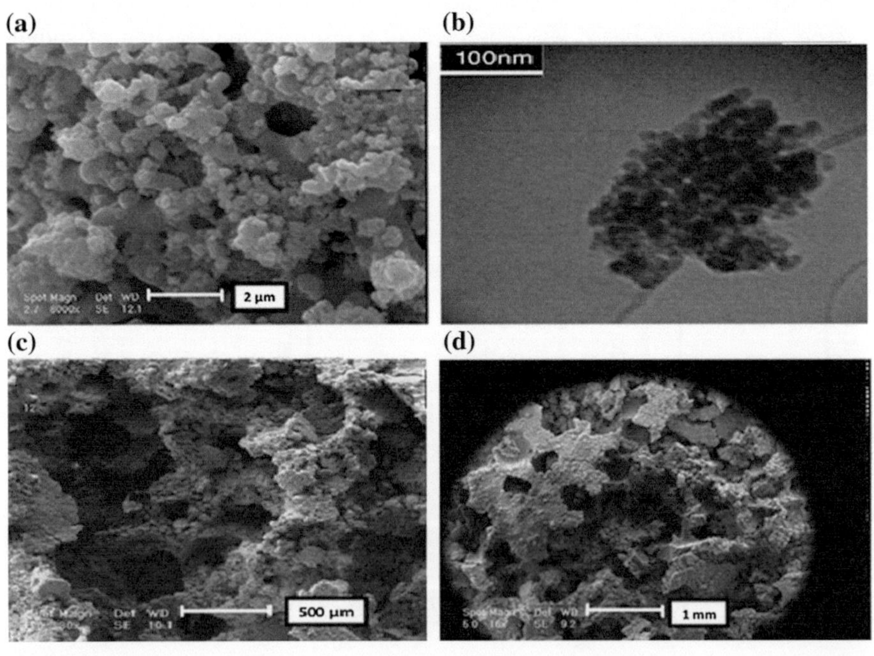

Fig. 2 **a** SEM images, **b** TEM images of hardystonite powder after 20 h ball mill followed by annealing at 1000 °C for 3 h, **c** and **d** SEM images of hardystonite scaffold with different magnification

Table 1 The mechanical and physical properties of hardystonite scaffold

Sample	Compression strength (MPa)	Compressive modulus (MPa)	Total porosity (%)	Interconnected porosity (%)	Size of porosity (μm)
Hardystonite	0.35	10.49	81	76	200–300

Based on the Archimedes' method [15], the interconnected and total porosity were measured to be 76 and 81%, respectively. As shown in Table 1, the compressive strength and modulus were measured to be 0.35 and 10.49 MPa, respectively. Wu et al. [10] prepared pure hardystonite scaffold with 86.94% porosity and pore size of 919.6 μm by foamy method. They reported the compressive strength of 0.06 MPa. Based on the compressive strength values of spongy bone (0.2–4 with about 70–90% porosity), the results of this study for hardystonite scaffold were close to spongy bone. A comparison between the results of this study with other reports indicated that using the space holder method can produce scaffolds with a higher strength and closer to the properties of spongy bone [17, 18].

In order to investigate the bioactivity of the hardystonite scaffolds, the simulated body fluid was used according to Bohner and Limatre [16]. Figure 3 shows SEM images of hardystonite scaffolds after 21 days of soaking in SBF. As can be seen, quasi-bone apatite was formed on the surface of the scaffolds which covered the entire surface and inside of the pores. As can be seen from the higher magnification images, apatite colonies with size of 1–3 μm with a wormy like morphology can be observed.

Figure 4a shows the pH changes of hardystonite scaffold up to 21 days of soaking in SBF. As shown, most of the pH changes occurred in the first few days. After 3 days the pH increased from 7.4 to 8.22. These pH changes are as a result of the breakdown of calcium and zinc ions from the scaffold structure after soaking in SBF. This ionic displacement (release of zinc and calcium ions from the scaffold into the solution and absorption of H + and breakage of the bonds results in the formation of silanol groups on the surface), caused the formation of a negative charge on the surface. Positive ions adsorbed to the surface and decreased the pH value in the solution (after about 8 days of immersion), and subsequently re-positively charged onto the scaffold surface and re-grouped due to the presence of positive charge on the surface. Calcium–phosphate groups are formed and eventually led to the formation of crystalline apatite on the surface [16]. Figure 4b shows the FTIR spectrum of the hardystonite scaffold before and after immersion in SBF after 21 days. The characteristic peaks at 1455 cm^{-1}, 499 cm^{-1} are related to Zn–O and Ca–O groups [19]. Also, the adsorbtion peaks at 617, 617, 681, 835, 911, 970, and 1010 cm^{-1} confirmed the presence of Si-O-Si groups.

The FTIR pattern of hardystonite scaffold after 21 days soaking in SBF showed the formation of apatite group on the surface of the scaffold. The characteristic peaks of calcium–phosphate groups were observed at 636, 918, 944, 971, and 1036 cm^{-1}. The peaks observed at 1621 and 474 cm^{-1} belong to the hydroxyl (H–O) group in the apatite structure. The peak of 1420 cm^{-1} also belongs to the carbonate

Fig. 3 SEM images of hardystonite scaffold after 21 days soaking in SBF with different magnification

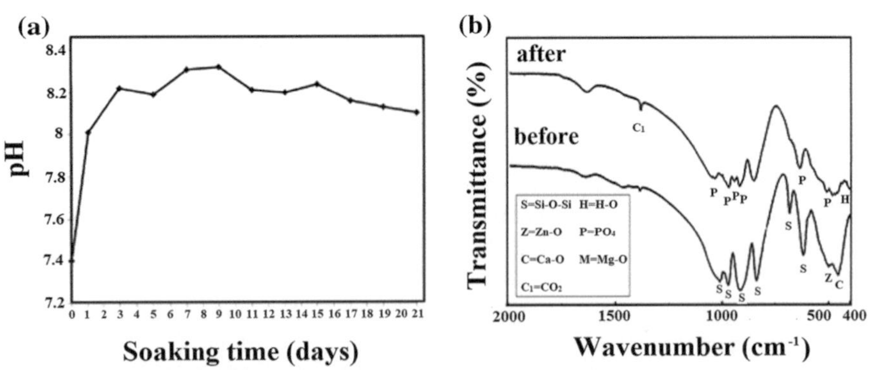

Fig. 4 **a** pH changes and **b** FTIR patterns of hardystonite scaffolds

group in the apatite structure [19]. The results of FTIR patterns were in a good agreement with the pH changes in SBF and the presence of apatite on the scaffold surface. Apatite colonies are also clearly identified in the SEM images [18, 19].

Conclusion

In this study, pure hardystonite powder was successfully fabricated by mechanochemical method and then pure hardystonite scaffold was produced using space holder technique with a pores size in the range of 200–300 μm. The average crystallite size of the powder and hardystonite scaffold were measured to be 28 and 79 nm, respectively. The compressive strength, compressive modulus, total and interconnected porosity values for this scaffold were 0.35 MPa, 10.49 MPa, 81%, and 76%, respectively. According to the results of the bioactivity test of hardystonite scaffold in SBF for 21 days, the presence of apatite was confirmed. Based on the SEM images the apatite covered the entire surface of the scaffold as well as the pores. The fabricated scaffold with appropriate pore size and structure is a good candidate for bone tissue engineering applications.

References

1. Burg KJL, Porter S, Kellam JF (2000) Biomaterial developments for bone tissue engineering. Biomaterials 21:2347–2359
2. Shirtliff VJ, Hench LL (2003) Bioactive materials for tissue engineering regeneration and repair. J Mater Sci 38:4697–4707
3. Roohani-Esfahani SI, Dunstan CR, Davies B, Pearce S, Williams R, Zreiqat H (2012) Repairing a critical-sized bone defect with highly porous modified and unmodified baghdadite scaffolds. Acta Biomater 8:4162–4172
4. Ghomi H, Emadi R, javanmard SH (2016) Fabrication and characterization of nanostructure diopside scaffolds using the space holder method: effect of different space holders and compaction pressures. Mater Des 91:193–200
5. Sadeghzade S, Emadi R, Ghomi H (2015) Mechanical alloying synthesis of forsterite-diopside nanocomposite powder for using in tissue engineering. Ceram Silik 59:1–5
6. Epple M (2018) Review of potential health risks associated with nanoscopic calcium phosphate. Acta Biomater 77:1–14
7. Hafezi M, Nezafati N, Nadernezhad A, Ghazanfari SMH, Sepantamehr M (2014) Bioinorganics in bioactive calcium silicate ceramics for bone tissue repair: bioactivity and biological properties. Ceram Sci Technol 5:1–12
8. Sadeghzade S, Emadi R, Tavangarian F (2016) Combustion assisted synthesis of hardystonite nanopowder. Ceramic Int 42:14656–14660
9. Wu C, Ramaswamy Y, Zreiqat H (2010) Porous diopside (CaMgSi2O6) scaffold: a promising bioactive material for bone tissue engineering. Acta Biomater 6:2237–2245
10. Wu C, Chang J, Zhai W (2005) A novel hardystonite bioceramic: preparation and characteristics. Ceram Int 31:27–31

11. Zreiqat H, Ramaswamy Y, Wu C, Paschalidis A, Lu Z, Birke O, Mcdonald M, Little D, Dunstan CR (2010) The incorporation of strontium and zinc into a calcium-silicon ceramic for bone tissue engineering. Biomaterials 31:3175–3184
12. Wang G, Lu Z, Dwarte D, Zreiqat H (2012) Porous scaffolds with tailored reactivity modulate in-vitro osteoblast responses. Mater Sci Eng, C 32:1818–1826
13. Sadeghzade S, Emadi R, Labbaf S (2016) Formation mechanism of nano-hardystonite powder prepared by mechanochemical synthesis. Adv Powder Technol 27(5):2238–2244
14. Gheisari H, Karamian E, Abdellahi M (2015) A novel hydroxyapatite- hardystonite nanocomposite ceramic. Ceram Int 41:5967–5975
15. Sadeghzade S, Emadi R, Tavangarian F, Naderi M (2017) Fabrication and evaluation of silica-based ceramic scaffolds for hard tissue engineering applications. Mater Sci Eng, C 71:431–438
16. Bohner M, Lemaitre J (2009) can bioactivity be tested in vitro with sbf solution? Biomaterials 30:2175–2179
17. Tavangarian F, Emadi R (2010) Mechanochemical synthesis of single phase nonocrystalline forsterite powder. Int J Mod Phys B 24:343–350
18. Sadeghzade S, Emadi R, Labbaf S (2017) Hardystonite-diopside nanocomposite scaffolds for bone tissue engineering applications. Mater Chem Phys 202:95–103
19. Soundrapandian C, Datta S, Kundu B, Basu D, Sa B (2010) Porous bioactive glass scaffolds for local drug delivery in osteomyelitis: development and in vitro characterization. Am Assoc Pharm Sci 11:1675–1683

The Hardystonite/PA66 Composite for Using as the Intervertebral Fusion Cage

Fariborz Tavangarian, Sorour Sadeghzade and Rahmatollah Emadi

Abstract The aim of this study was to design and evaluate nano-hardystonite/polyamide 66 composites with close mechanical properties to the trabecular bone to prevent the stress shielding phenomenon for bone tissue engineering applications. This composite can be used as an intervertebral fusion cage to perform spinal fusion between vertebrae in the lumbar spine. The pure nano-hardystonite powder was fabricated by combustion method at 900 °C following by 5 h ball mill. The nano-hardystonite/Polyamide 66 and nano-hydroxyapatite/Polyamide 66 cage were prepared by injection molding method to compare the mechanical and biological properties. In this end, the X-ray diffraction (XRD) and scanning electron microscopy (SEM) were utilized to characterize the prepared powder and cage samples. Based on the results, the addition of 30% nano-hardystonite improved both mechanical and bioactivity properties. The optimum hardystonite/PA66 samples revealed the compressive strength and elastic modulus of 69.19 ± 0.89 MPa and 2.56 ± 0.5 GPa, respectively, compared to 53.45 ± 1.2 MPa and 3.45 ± 0.3 GPa in hydroxyapatite/PA66 sample, respectively. In addition, observation of the superior apatite formation ability of hardystonite/PA66 compared to hydroxyapatite/PA66 indicated that it can be used as a spinal vertebrae replacement material.

Keywords Cage · Polyamide · Hydroxyapatite · Hardystonite · Spinal vertebrae

F. Tavangarian (✉) · S. Sadeghzade
Mechanical Engineering Program, School of Science, Engineering and Technology, Pennsylvania State University, Harrisburg, Middletown, PA 17057, USA
e-mail: f_tavangarian@yahoo.com

S. Sadeghzade · R. Emadi
Materials Research Group, Department of Materials Engineering, Isfahan University of Technology, 84156-83111 Isfahan, Iran

B. Li et al. (eds.), *Advances in Powder and Ceramic Materials Science*, The Minerals, Metals & Materials Series,
https://doi.org/10.1007/978-3-030-36552-3_16

Introduction

Back pain is one of the most common reasons people go to the doctor or miss work, and it is a leading cause of disability worldwide [1, 2]. Surgery may be considered for severe lower back pain that does not get better after a 6–12 week course of nonsurgical treatments. The spine surgery is one of the most challenging operation. Fusion surgery basically removes the soft tissues between two or more adjacent vertebral bones and replaces them with implant. This procedure enables the bones to grow together over time typically 6–12 months and fuse into one long bone to stabilize and eliminate motion at those spinal segments [3, 4]. Using the implants instead of autografting and allografting is due to their limitation such as lack of donors, increasing infection, require double operations, osteomyelitis, and immune response [5, 6]. Hence, synthetic devices are the best choice as alternative methods to encourage the bone tissue growth [7, 8]. The cage devices are a new-generation of spinal vertebrae implants which can be placed as a new device to perform the spinal fusion. The cage was made of metal, polymer, ceramic, and composite of these materials in recent decades [9]. Using the metal as intervertebral spinal cage was reported in the previous studies. For example, titanium mesh cage showed the high fusion rates. But the disadvantages of metals such stress shielding, high incidence, rate of corrosion, and creating problems in radiography procedure are the main reasons that limited their usage as an ideal cage [8]. Therefore, polymer and ceramic materials can be suggested as new materials to fabricate the cage devices. One of the best polymer with excellent biocompatibility is Polyamide (PA). In fact, this synthetic polymer with the chemical networks and active group close to collagen has been attractive for use in medical applications. But the main disadvantage of PA which limited its biomedical applications is the fact that it is a bio-inert polymer [9, 10]. Hardystonite and Hydroxyapatite (HA) are well known as calcium silicate and calcium phosphate group with high bioactivity and biocompatibility. Although, the biological capability of hydroxyapatite has been reported to be higher than that of hardystonite, the poor mechanical properties of hydroxyapatite have limited its application in medical applications [11–13]. Using a composite of polyamide and hardystonite for designing the intervertebral fusion cage with improved mechanical and biological properties is the main aim of this study. Therefore, the in vitro bioactivity and compressive strength of hardystonite/Polyamide composite cage were investigated. In addition, the mechanical and biological properties of hardystonite/PA66 cage were compared with hydroxyapatite/PA66 cage.

Materials and Method

Hardystonite and Hydroxyapatite Powder Fabrication

The Calcium nitrate tetra- hydrate, zinc nitrate hexahydrate and tetraethyl-orthosilicate were used as initial materials to produce the nano-hardystonite powder with the molar ratio of Zn:Ca:Si = 1:2:2. In this end first, TEOS was hydrolyzed in deionized water (molar ratio of 1:8) for 30 min. Then the zinc nitrate and calcium nitrate powders were mixed with hydrolyzed TEOS for 30 min. Then the mix of PVA/sucrose with the metal molar ratio of 2:0.4:1 added to the aqueous as the fuel agent. After that, the solution stirred for 24 h and the gel was put on oven for 1 day. Finally, the dried gel was ball milled for 5 h and the final powder was annealed for 3 h at 950 °C.

The bovine bone was used to produce the hydroxyapatite powder. First, the bone boiled for 5 h in distilled water to remove the visible tissue. Then, it burned at 1000 °C for 3 h and the matter was ball milled for 10 h in zirconia vial with five 2 cm zirconia balls. The ball/powder weight ratio was 10:1 and disc rotational speed was set 250 rpm.

Cage Fabrication

The 30% hardystonite and hydroxyapatite ceramic powders were added to PA66 cage. The cage was prepared by injection molding method.

Characterization of Powder and Composite

The X-ray diffraction (XRD, X'pert philips) with CuKα (λ = 0.154 nm) radiation at 40 kV and 30 mA was utilized to characterize the phase transformation of hardystonite. The XRD patterns were recorded at the 2Θ range of 20–70° with 0.05° step size and 1 s time per step.

In order to evaluate the apatite form ability of fabricated cage the samples were soaked in simulated body fluid (SBF) for 28 days. Scanning electron microscopy (SEM, Philips XL30 at an acceleration voltage of 30 kV) was used to evaluate the bioactivity of produced cage in SBF. The universal testing machine (Hounsfield: H25KS) was used to measure the compressive strength and modulus of the produced cage.

Results and Discussion

The XRD pattern of hardystonite and hydroxyapatite powder was shown in Fig. 1. As can be seen the characteristic peaks of hardystonite (JCPDS 1-075-0916) (Fig. 1b) and hydroxyapatite (JCPDS 01-086-0740) (Fig. 1d) phases were observed which confirmed the formation of pure powders. The morphology of hardystonite (Fig. 1a) and hydroxyapatite (Fig. 1c) powder were evaluated by SEM micrograph. The wormy shape of hardystonite particles with uniform distribution can be observed. Also, the mean particle size of hardystonite powder was found to be 220 ± 4 nm.

Also, the SEM micrograph of hydroxyapatite powder shows the irregular agglomerated particles in the range size of 300–400 nm. It is shown that the effect of fabrication method to the particles shape and size.

Improving some features such as reduced sintering temperature and high contact area reduced the diffusion paths which made the nano-crystalline ceramics more attractive than micro particle ceramics [14, 15]. Hence, the nano-size ceramic is preferred to be used as tissue engineering materials. The mechanical properties of the prepared HT30%/PA66 and HA30%/PA66 cage, in this study, and trabecular and cortical bone are demonstrated in Table 1.

Based on the mechanical results which can be observed from Table 1, the type of ceramic additive and the design of cage (cage with and without hole, number of holes) are the main factors that can affect the mechanical properties of the produced cages. The results (Table 1) demonstrated that the addition of 30% hardystonite in the matrix of PA66 (the cage without hole) caused the highest compressive strength

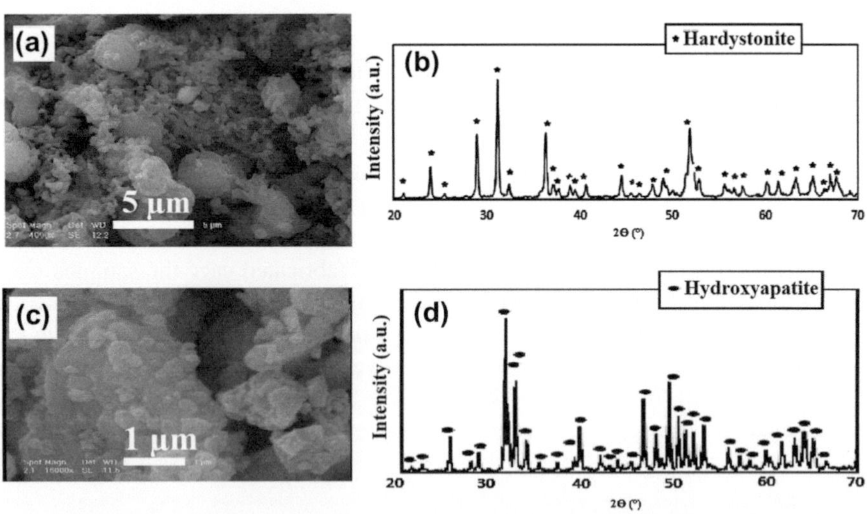

Fig. 1 **a** SEM micrograph, **b** XRD pattern of hardystonite powder, **c** SEM micrograph, **d** XRD pattern of hydroxyapatite powder

Table 1 Mechanical properties of produced cage, spongy and cortical bone (mean ± standard deviation)

Samples	Compressive strength (MPa)	Elastic modulus (GPa)	Number of hole
PA66	57.27 ± 1.5	1.55 ± 0.8	2
HA30%/PA66	73.51 ± 1.32	7.21 ± 0.34	No
HA30%/PA66-2	53.45 ± 1.2	3.45 ± 0.3	2
HT30%/PA66	110.48 ± 1	10.21 ± 0.81	No
HT30%/PA66-2	69.19 ± 0.89	2.56 ± 0.5	2
Spongy bone	4–12 [15]	0.1–1.1 [15]	No
Cortical bone	130–180 [15]	7–30 [15]	No

and modulus compared to cage with 30% HA with the same design. As can be seen from the SEM micrograph (Fig. 2) of the surface of HT30%/PA66, the high interaction between the hardystonite and PA66 matrix can be observed.

This can be ascribed to the electrostatic forces and hydrogen bonds existing between the Ca, Si, and Zn ions and −C−O and −NH − functional groups in PA66. Additionally, coordination bonds were formed between the −COO group of PA66 and Ca^{2+}, Si, and Zn ions of hardystonite [9, 10]. Therefore, the improved mechanical properties can be as a result of this chemical interaction. The elastic modulus of implants' devices is very important and should be close to the bone tissue. The higher or lower Young modulus compared to desired tissue can cause the stress shielding phenomena [12]. Therefore, using the materials with elastic modulus close to that of human bone is desirable. As can be seen from Table 1, the HT30%/PA66 showed the higher modulus compared to spongy bone. The different designing in cage could improve the modulus and make its mechanical properties closer to that of spongy bone. In this case, both compressive strength and modulus

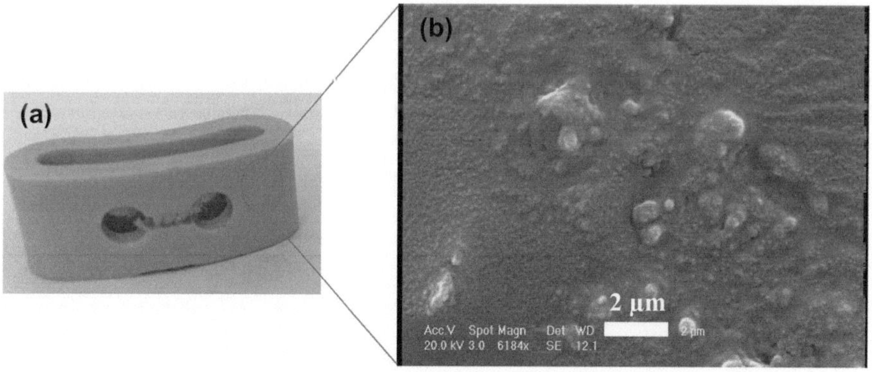

Fig. 2 **a** HT30%/PA66 cage with hollow elliptical shape and 27 mm in larger outer diameter and 21 mm in inner diameter, **b** SEM images of the surface of HT30%/PA66 cage

Fig. 3 SEM micrograph of **a** PA66, **b** HA30%/PA66 and **c** HT30%/PA66 after 28 days soaking in SBF

of HT30%/PA66 with two holes were found to be 69.19 ± 0.89 MPa and 2.56 ± 0.5 GPa, respectively, which was in the range of compressive strength and modulus of spongy bone.

The SEM micrograph of HT30%/PA66 and HA30%/PA66 were demonstrated in Fig. 3 after 28 days soaking in SBF.

As can be seen, more precipitated white particles can be observed on the surface of cage with 30%HT compared to HA which was not expected. The main reason for this phenomenon can be due to the size of ceramic particles. In fact, the nano particles can increase the bioactivity in the implant, while the ceramic with micron size showed a lower apatite formation ability. Also, it should be noted that the addition of hydroxyapatite and hardystonite to PA66 compared to PA66 cage alone after 28 day soaking in SBF lead to better apatite formation ability. It should be mentioned that PA66 is known as bio-inert polymer which does not show any apatite formation ability and is considered as a bio-inert polymer.

Therefore, the HT30%/PA66 cage with two holes improved the mechanical properties and bioactivity of the implants and can open a bright horizon to treat the problem of degenerative disc disease.

Conclusion

In this study, the HT30%/PA66 cage was successfully fabricated by injection molding method to develop a new intervertebral fusion cage with improved mechanical properties and bioactivity. The results showed that the addition of 30wt% hardystonite with 220 nm mean particle size to polyamide matrix caused an improvement in the compressive strength and elastic modulus (69.19 $\pm$ 0.89 MPa and 2.56 $\pm$ 0.5 GPa) compared to HA30%/PA66 which was closer to that of spongy bone, respectively. Moreover, the improvement in bioactivity of HT30%/PA66 was observed compared to HA30%/PA66 and PA66.

References

1. Yang X, Song Y, Liu L, Liu H, Zeng J, Pei F (2012) Anterior reconstruction with nano-hydroxyapatite/polyamide.e-66Cage After Thoracic and Lumbar Corpectomy. Orthopeedics 35:66–73
2. Roohani-Esfahani SI, Chen Y, Shi J, Zreiqat H (2013) Fabrication and characterization of a new, strong and bioactive ceramic scaffold for bone regeneration. Mater Lett 107:378–381
3. Zhao Z, Jiang D, Ou Y, Tang K, Luo X, Quan Z (2012) A hollow cylindrical nano-hydroxyapatite/polyamide composite strut for cervical reconstruction after cervical corpectomy. J Clin Neurosci 19:536–540
4. Xiong Y, Ren C, Zhang B, Yang H, Lang Y, Min L, Zhang W, Pei F, Yan Y, Li H, Mo A, Tu C, Duan H (2014) Analyzing the behavior of a porous nano-hydroxyapatite/poly. Amide 66 (n-HA/PA66) composite for healing of bone defects. Int J Nanomedicine 9:485–494
5. Roohani-Esfahani SI, Dunstan CR, Davies B, Pearce S, Williams R, Zreiqat H (2012) Repairing a critical-sized bone defect with highly porous modified and unmodified baghdadite scaffolds. Acta Biomater 8:4162–4172
6. Hutmacher DW, Schantz JT, CX L, KC T, TC L (2007) State of the art and future directions of scaffold-based bone engineering from a biomaterials perspective. J Bone Joint Surg Am 1:60–245
7. Ht H, Me M, Rt H (2003) Complications of multilevel cervical corpectomies and reconstruction with titanium cages and anterior plating. J Spinal Disord 16:1–9
8. Thalgott JS, Xiongsheng C, Giuffre JM (2003) Single stage anterior cervical reconstruction with titanium mesh cages, local bone graft, and anterior plating. Spine J 3:294–300
9. Zhanga X, Zhanga Y, Zhangb X, Wanga Y, Wanga J, Lua M, Li H (2015) Mechanical properties and cytocompatibility of carbon fibre reinforced nano-hydroxyapatite/polyamide66 ternary biocomposite. J Mech Behave Biomed Mater 42:267–273
10. Xu Q, Lu H, Zhang J, Lu G, Deng Z, Mo A (2010) Tissue engineering scaffold material of porous nanohydroxyapatite/polyamide 66. Int J Nanomedicine 5:331–335
11. Sadeghzade S, Emadi R, Labbaf S (2016) Formation mechanism of nano-hardystonite powder prepared by mechanochemical synthesis. Adv Powder Technol 27:2238–2244
12. Sadeghzade S, Emadi R, Tavangarian F, Naderi M (2017) Fabrication and evaluation of silica-based ceramic scaffolds for hard tissue engineering applications. Mater Sci Eng C 71:431–438

13. Ramaswamy Y, Wu C, Zhou H, Zreiqat H (2008) Biological response of human bone cells to zinc-modified Ca–Si-based ceramics. Acta Biomater 4:1487–1497
14. Seiler JG, Johnson J (2000) Iliac crest autogenous bone grafting: donor site complications. J South Orthop Assoc 9:91–97
15. Sadeghzade S, Shamoradi F, Emadi R, Tavangarian F (2017) Fabrication and characterization of baghdadite nanostructured scaffolds by space holder method. J Mech Behave Biomed Mater 68:1–7

Author Index

B. Li et al. (eds.), *Advances in Powder and Ceramic Materials Science*,
The Minerals, Metals & Materials Series,
https://doi.org/10.1007/978-3-030-36552-3

Subject Index

B. Li et al. (eds.), *Advances in Powder and Ceramic Materials Science*,
The Minerals, Metals & Materials Series,
https://doi.org/10.1007/978-3-030-36552-3

MIX
Papier aus verantwortungsvollen Quellen
Paper from responsible sources
FSC® C105338

If you have any concerns about our products,
you can contact us on
ProductSafety@springernature.com

In case Publisher is established outside the EU,
the EU authorized representative is:
Springer Nature Customer Service Center GmbH
Europaplatz 3, 69115 Heidelberg, Germany

Printed by Libri Plureos GmbH
in Hamburg, Germany